P9-DMY-220

ALSO BY DAVID KUSHNER

Alligator Candy: A Memoir

The Bones of Marianna: A Reform School, a Terrible Secret, and a Hundred-Year Fight for Justice

Jacked: The Outlaw Story of Grand Theft Auto

Levittown: Two Families, One Tycoon, and the Fight for Civil Rights in America's Legendary Suburb

Jonny Magic & the Card Shark Kids: How a Gang of Geeks Beat the Odds and Stormed Las Vegas

Masters of Doom: How Two Guys Created an Empire and Transformed Pop Culture

GRAPHIC NOVEL:

Rise of the Dungeon Master: Gary Gygax and the Creation of D&D (illustrated by Koren Shadmi)

AUDIO BOOKS:

Prepare to Meet Thy Doom: And More True Gaming Stories

The World's Most Dangerous Geek: And More True Hacking Stories

A GENIUS, A CON MAN,

and the

SECRET HISTORY

of the INTERNET'S RISE

the PLAYERS BALL

DAVID KUSHNER

Simon & Schuster
New York London Toronto Sydney New Delhi

Simon & Schuster
1230 Avenue of the Americas
New York, NY 10020

First Simon & Schuster hardcover edition April 2019

SIMON & SCHUSTER and colophon are registered trademarks of Simon & Schuster, Inc.

For information about special discounts for bulk purchases, please contact Simon & Schuster Special Sales at 1-866-506-1949 or business@simonandschuster.com.

The Simon & Schuster Speakers Bureau can bring authors to your live event. For more information or to book an event contact the Simon & Schuster Speakers Bureau at 1-866-248-3049 or visit our website at www.simonspeakers.com.

Interior design by Carly Loman

Manufactured in the United States of America

10 9 8 7 6 5 4 3 2 1

Library of Congress Cataloging-in-Publication Data

Names: Kushner, David, author.
Title: The players ball : a genius, a con man, and the secret history of the Internet's rise / David Kushner.
Description: New York : Simon and Schuster, [2019]
Identifiers: LCCN 2018032968| ISBN 9781501122149 (hardcover) | ISBN 9781501122156 (trade pbk.)
Subjects: LCSH: Internet industry—United States—History. | Internet domain names. | Dating services—United States. | Online dating—United States. | Swindlers and swindling—United States.
Classification: LCC HD9696.8.U62 K87 2019 | DDC 384.30973—dc23 LC record available at https://lccn.loc.gov/2018032968

ISBN: 978-1-5011-2214-9
ISBN: 978-1-5011-2216-3 (ebook)

For Sue, my perfect match

CONTENTS

Live in fragments no longer. Only connect, and the beast and the monk, robbed of the isolation that is life to either, will die.

—E. M. Forster, *Howards End*

KATE (*SINGING*): The internet is really, really great!

TREKKIE MONSTER: For porn.

—*Avenue Q: The Musical*

the PLAYERS BALL

CHAPTER 1

THE WORLD WILD WEST

No one expected things to get so dirty.

It was just a local election, and a seemingly inconsequential one at that, a seat on the Santa Clara Valley Water District Board of Directors. The seven-person board manages the water system and flood control for the 1.9 million residents of the California county, which includes Silicon Valley: patching dams, overseeing water treatment plants, stocking sandbags when creeks overflow, and so on. It's a noble but unglamorous public service compared to the jetset lives of the tech titans in town. The only residents who usually bother to attend the public meetings are a handful of retirees, and the homeless woman who often sleeps in the back.

But, in 2014, voters were more interested than ever. "Water district elections are usually low-key—if not boring—affairs," the *San Jose Mercury News* reported. "Not this year." The two-man race had become the most vicious, and confounding, in the Santa Clara Valley Water District's eighty-five-year history. There'd been allegations of corruption, sexual depravity, scandalous lies. For reasons no one could gather, Gary Kremen, a heavyset, disheveled, fifty-one-year-old dot-com multimillionaire, Deadhead, Stanford MBA, and self-described "kook," shelled out $408,492—largely from his own pocket—to beat incumbent Brian Schmidt, an earnest forty-seven-year-old environmental attorney who'd devoted his career to the cause. "Why is he spending so much?" Schmidt told the *Mercury News*. "I don't know what to say."

The water of Silicon Valley pumped through the heart of Schmidt, who came off like an Eagle Scout. He'd earned his law degree from Stanford, labored locally as an environmental attorney, and blogged about the box turtles he saw while cleaning up the coast. For the past four years, he had proudly served on the Water District board. His fight for the potable reuse of recycled waste water, which could supply half the county's drinking supply, helped him earn multiple media endorsements. But with only a few days until the election, and his money (and dreams) running out, he finally had enough of Kremen's Animal House behavior.

One morning in late October, Schmidt bicycled alone to a dried-up pond in the woods near his home in Palo Alto to shoot a last-ditch campaign video for YouTube. Slowly panning his camera across the landscape, he filmed the muddy field limned with dying brown trees. "What you're seeing around you is the effect of the California drought," he narrated solemnly. Then Schmidt set the camera in place, and stepped in front of it to tape himself. He was prematurely gray but boyish, and wore a blue "Re-Elect Brian Schmidt" T-shirt. "I am kind of proud to say I am now a target of a negative mailer," he said.

Schmidt approached the camera and held up the cover of said mailer: a custom greeting card that mocked his recycled waste water plan. "BRIAN SCHMIDT wants to get our drinking water from OUR TOILETS," the card read.

"It's a picture of me—next to a toilet," Schmidt explained. "It claims not to be from my opponent, but you can take that for whatever you want to take it for. This is a very expensive thing, where my opponent has put a lot of money into the race."

As Schmidt opened the greeting card, it played an audio snippet from one of his stump speeches: "I'm advocating treatment of waste water to drinkable levels." Then a woman's voice came on: "Brian Schmidt wants my family to drink water from a toilet? Ewww!" she said. "Say no to toilet water! Say no to Brian Schmidt for Santa Clara Valley Water District."

As his camera rolled, Schmidt stepped back into focus with the beleaguered expression of a science teacher who'd sat on one too many whoopee cushions. Pedantically, he explained that people were already drinking recycled waste water from Singapore to the International Space Station. "This is astronaut water we're talking about," he went on. "It's healthy enough for them, it's healthy for the rest of us." He appealed to the brainiacs in town to give him, and his astronaut water, a chance. "This is Silicon Valley," he said, as he concluded recording the video he later posted online. "This is a highly educated area. We understand what we can do with technology."

But few understood technology better than the highly educated man so curiously obsessed with beating him, Gary Kremen. Silicon Valley had seen its share of iconoclastic visionaries—Steve Jobs, Bill Gates, Mark Zuckerberg—but none like Kremen. Though largely unfamiliar to the outside world, he was among the most prescient entrepreneurs in the history of the internet. In business, and in life, true visionaries not only have the foresight to find the next frontier, but the confidence to bet on it. Kremen was legendary among those in the know for his uncanny cocktail of both. As veteran technology investor Ron Posner put it, "He's very energetic, very creative, very smart—and never gives up."

Kremen is the father of online dating. In 1993, he founded what was essentially the first—now biggest—dating site, Match.com. Despite about only 5 percent of Americans being online around that time, Kremen brashly told a skeptical TV reporter in 1995 that his invention was going to change the world. "Match.com will bring more love to the planet than anything since Jesus Christ," the then thirty-one-year-old declared in his nasally toned Chicago accent. The fact that this prediction was coming from some Belushi in a stained tie-dye T-shirt sprawled on a red bean bag made it all the more dubious.

But as Kremen would prove time and time again, his hunch was right. Match.com became an international phenomenon, spreading to more than twenty-five countries in eight languages with more than 42 million members, and becoming the basis of today's $2 billion online dating industry. The company Kremen started with a $2,500 credit card loan now has a value of $3.5 billion. At a time when most businesspeople barely understood, let alone paid attention to, the internet, Kremen was among the first to figure out how to make money online. Even more radically, he transformed the way people meet and marry in the digital age. As he wrote with characteristic humor on his Water District campaign website bio, "I am responsible for over 1,000,000 babies!"

But according to his detractors, he was responsible for much that was wrong with the internet too. The genius of love was also the sultan of sex, specifically Sex.com, one of the most notorious websites ever online. And it was his epic battle over Sex.com that made him most legendary of all.

It started in the early 1990s, before he pioneered online dating, when Kremen envisioned another new frontier. He thought that domain names—the dot-coms and other addresses that signify ownership online—would one day have the value of real property as people learned how to build businesses on the net. Because domains were deemed worthless at the time, they were essentially free to register. So Kremen gobbled up dozens—Jobs.com, Housing.com, Autos.com, Match.com, and the like—to later monetize. It was the online equivalent of coming to America and staking claims across the country. Though he had no intention of becoming a pornographer, in 1994 he registered Sex.com too, thinking it could become a health and wellness education site.

But while he had been busy with Match.com, someone named Stephen Michael Cohen had somehow stolen the rights to the Sex.com domain—and transformed it into one of the most profitable websites, earning millions of dollars a month. Kremen wanted his site back, and he wanted to get the money Cohen had earned through stealing it. What followed was an epic rivalry that established many of the rules that enable online commerce today. As Kremen's esteemed lawyer James Wagstaffe said, "This case established the precedent that a domain name is property—property that can be stolen."

On one side was Kremen. A devoted father of two, he epitomized the entrepreneurial spirit of the new Silicon Valley in his quest to bring love and connection to the world. On the other side was his evil genius twin, Cohen, a stout, balding, forty-eight-year-old convicted felon from Los Angeles. As a con man, Cohen was among the best. He impersonated judges and lawyers, ran swingers clubs, left trails of bad checks, and scammed investors after promising to build a Fantasy Island of sex in the desert outside Las Vegas. A lifelong computer enthusiast, Cohen insisted that, despite Kremen's claims, he was the real originator of internet dating, and the rightful owner of Sex.com, long before Kremen came on the scene. "He says he invented online sex. That's the most ridiculous thing," as Cohen said. "I started the sucking and fucking online!

I invented that!" The *Los Angeles Times* called Cohen "one of the most successful entrepreneurs of the Internet Age."

The battle of Cohen and Kremen became highly personal, and picaresque. It ran from the boardrooms of Silicon Valley to the bordellos of Mexico. There were porn stars and programmers, billionaires and brainiacs, goons and gangsters. There was even a gunfight on the streets of Tijuana, according to Cohen at least.

But there was more than fame and fortune at stake. The war for Sex.com represents an essential, but overlooked, chapter behind one of the greatest inventions of our time: the internet. In the public imagination, the computer revolution is bookended by two main stories, the rise of Apple and personal computing in the 1980s, and the proliferation of Facebook and social media in the 2000s. But in between those years is a lost era of innovation that's crucial to understanding the rapidly evolving world of today and tomorrow: the internet gold rush of the 1990s. As in the American Frontier of the 1800s, the early settlers of the net fought to stake their claim and make their millions. They established the systems that defined the world to come. But none of this would have been possible without the fuel that made the internet what it is today, love and sex.

And so in the fall of 2014, it seemed all the more confounding why a mogul with Kremen's history would be investing so much time and money to win the Water District race. Kremen claimed it was part of his burgeoning interest in sustainability. One of the recent start-ups he founded, a solar energy firm called Clean Power Finance, had received $75 million in investments from Google and others. Brian Schmidt, his beleaguered opponent, suspected there had to be some other reason—he just couldn't figure it out. "It's tricky to try to get at somebody's deeper motivations," as Schmidt told the *Mercury News*, "but it is really concerning."

With every story, there's a road. Kremen's led back to the Wild West days of the world wide web, and the fight for the future that left him—and Cohen—reeling to this day. "Believe whatever you want," as Cohen said late one night not long ago, "I don't care. I don't drive a Rolls-Royce anymore. I don't own airplanes anymore. I don't have boats anymore. Because none of that shit means anything to me anymore."

CHAPTER 2

THE GENIUS

Gary Kremen nervously lingered inside his grandfather Manny's barbershop, ready to bust his move. It was the late 1970s in Lincolnwood, Illinois, a working-class suburb an hour northwest of Chicago near Skokie, but Kremen, a chubby teen with a nest of dark hair, wasn't here for a trim. Just as his grandfather turned his head for his broom, Kremen slid something under his shirt and sprinted out the door.

Hopping on his bike, he pedaled quickly away through his run-down neighborhood, which had earned the nickname "the Toilet." He zipped past the baseball team gorging at Lou Malnati's Pizza, the kids playing putt-putt at the Bunny Hutch Miniature Golf, the solemn commuters on the 210 bus to Chicago. He turned off the main road, as the brick buildings turned to modest homes with dads washing their cars and kids shooting hoops outside, a left here, a right there, until he peeled up the driveway of a pale-pink thousand-square-foot ranch house on a dead-end street by the train track, tossed his bike to the ground, lumbered inside over the mustard-yellow and brown thick shag carpet, past the flowered couches covered in plastic, down the steps into the wood-paneled basement, locking the door behind him, reaching into his shirt, and slipping out his coveted bounty: a *Playboy* magazine.

But the magazine wasn't for him, at least not after he was done with it. Kremen was young, but he was in business. And his business was selling the boys of Lincolnwood the hard-to-get stuff they wanted. When they wanted Hubba Bubba, the delectably soft cubes of bubble gum that had recently hit the market and were flying off the shelves, he bought up cases with his lawn mowing and snow shoveling money, then pawned them, piece by piece, to kids at four times the price he paid for them. With his profit, he diversified into baseball cards, dealing to the Cubs fanatics on the schoolyard.

But while candy fads and baseball stars came and went, there was one thing that never went out of style—porn magazines—and those were among the hardest things to get of all. The technology didn't exist yet to meet the illicit demand. There was no internet, or even cable TV. The only way for kids to get dirty pictures was to steal them, or hand over money to the one guy in town who had the nerve, if not the entrepreneurial ambition, to do it for them, Kremen. It wasn't just the money that motivated him. It was something deeper, the same thing that drove any successful person: the conviction that a new frontier beyond the Toilet awaited, where his wildest dreams could come true, if only he could find them. But he wasn't kidding himself. Getting there wouldn't be easy. He wasn't handsome or athletic or rich. If he was going to get there, he'd have to do it the only way a young nerd from the suburbs could do it at the time, by using his brains.

Fortunately, brains were a given in the Kremen household. His parents—Norman and Harriett—had used their own to make a comfortable life for Gary and his rebellious younger sister, Julie.

And, like Gary, they had risen up by their own bootstraps. Norman, the son of a newspaper distributor and homemaker, had put himself through college, and earned a master's in education, as had Harriett, the daughter of a barber and manicurist. After meeting on a blind date, his mom taught high school accounting, and his dad, who started teaching driver's education, worked his way up to becoming an elementary school principal.

Gifted and geeky, they ran their home like their own mental gymnasium, pushing their kids to engineer, explore, and compete. Harriett, a math whiz, encouraged Gary's interest in math and science, getting him to compete in science fairs and mathlete clubs. Norman, a ham radio hobbyist, nurtured Gary's interest in engineering. They built their own radio together, and installed a thirty-foot ham radio tower in their backyard. To get his own license, Kremen stayed up all night studying the requisite Morse code.

Gary and his father spent long afternoons in the basement, communicating to other ham operators around the world, putting orange thumbtacks on a map for each country they reached. Kremen marveled at the wonder of the world waiting for him out there, a world he could teach over the airwaves. It felt like a magical extension of their family road trips—when they'd pile into the powder blue station wagon and hit the road to drive to every state but Hawaii. And now he could just go in the basement and connect with people everywhere.

When Kremen was about twelve, in the mid-1970s, they built their own computer and won the science fair. With personal computers not yet on the market, the only way to have such a machine was to cobble one together from a kit. Kremen's device looked like something from *Star Trek*, a giant board with throw switches and lights that could add numbers. For a kid who devoured science fiction novels as ravenously as Chicago hot dogs, the computer fired his imagination like nothing else. When his friend showed him a computer running Microsoft BASIC language—a machine that could be programmed by anyone—he had one thought: *This is the fucking future.*

The computer wasn't just a taste of the next frontier, it was a means of empowerment—and a way to compete. Kremen joined the nascent computer club at high school, working for hours just to calculate pi to the furthest digit, and try to break the world record. When the group built their own electric motors, Kremen's was ugly, but functional. It was also, as his friend Mark Zissman observed, the only one that seemed to have been made by a kid without the help of a parent. The better Kremen did in the club against the other kids, the more access he'd get to the machines, and Kremen soon had what was considered "rock star" privileges—with more time and access than anyone. Classmate Neal Gussis recalled that Kremen was the kind of kid whom teachers either loved or hated.

Kremen's newfound swagger had been manifesting in delinquent ways: smoking pot, lighting fireworks, hopping on trains as they chugged past his cul-de-sac. The computer power didn't just bolster his ego and bragging rights, it gave him a means for revenge. An anti-Semitic teacher had been coming on to girls, telling them, "Here's a dime, call me when you're eighteen." One afternoon, Kremen struck back, using his access to get on the teacher's computer. He then secretly typed out a message calling his teacher a "fuckhead"—and had it print out on all the student homework assignments. He got some laughs, and two days out of school suspension.

The infraction didn't help him at home, where he was increasingly butting heads with his domineering mother. No matter what he did, he felt, it was never good enough for her. His room wasn't clean enough, his grades weren't high enough, and, as she reminded him again and again, he wasn't keeping pace with the other bright Jewish boys in the highly competitive town. And there was no one she reminded him about more than a kid named Sheldon—the mere mention of his name made Kremen's stomach churn.

If Kremen scored a 95 on a test, his mother would tell him Sheldon had scored 97. Kremen went around in stained T-shirts and faded jeans; Sheldon wore button-up shirts and slacks. Sheldon had even won the Golden Apple award, the highest honor a teacher could bestow on a student. And yet as much as Kremen resented Sheldon's success, he wanted to be him. And since there was no way, in the eyes of his mother, he could ever live up to the glory of Sheldon, Kremen acted out.

Kremen egged Sheldon's house. He sent him porn subscriptions. And, late one night, he snuck onto Sheldon's lawn with a can of Glade air freshener. No one saw him in the shadows as he furtively sprayed the lawn in a design of his intent. At first, his handiwork would be imperceptible. But, in the days to come, it would magically appear. And then one delectable morning, Sheldon traipsed outside with his perfectly parted hair and starched shirt to find a message waiting for him in the lawn where the Glade had achieved its toxic effect, killing the grass with which it had come into contact and leaving his message etched in the lawn.

"Fuck you, Sheldon," it read.

Early adopters like Kremen had to work hard to get access to computers in the early 1980s. Personal computers were still new and expensive. Apple had introduced the first one in 1977, and IBM's first PCs, which came in 1981, cost at least $1,500, more than most could afford. The only way for Kremen to get his fingers on a machine was at college.

While pursuing two degrees—one in electrical engineering and the other in computer science—at Northwestern, he got a job in the school administration's computer room, which was filled with the large mainframe machines of the time. Kremen had to wait for the other nerds to get off the terminals to give him a chance. Once on, Kremen was happy to feed the paper punch cards—which contained the computer code—into the machines, amazed by how the programs came to life.

Light-years ahead of the other students, he'd get bored and distracted in class, putting off his assignments until way after the due date—then backdating them on the computer to make it look as if he had turned them in on time. When other students wanted to know their grades, he sold them access—taking money to log on to the systems and give them the information. It wasn't even about the money so much as the power, as he later put it, "just to show how important I am." It was nice while it lasted—until a student turned him in, and Kremen got fired.

But the bug was firmly planted in his brain, and Kremen, with his skills and grades, had his pick of places to go after graduating cum laude to get his geek on. These were the Reagan years, and with the defense buildup, demand was high for computer engineers. Kremen took a job with Aerospace Corporation, a defense nonprofit outside Los Angeles, where his job was to evaluate the cost of software for defense projects and early network security. But, really, what appealed to him was something more primal: the opportunity to spend 1985, as he later put it, "playing on the nascent internet."

For Kremen, the internet felt like the natural evolution of ham radio, except that instead of going down to his father's basement he could sit at this machine. And it was all still younger than even him. The roots of the internet started in 1969, when the Advanced Research Projects Agency, the part of the United States Department of Defense responsible for technology research, launched the first online network, known as ARPANET. The purpose was to provide an emergency means for government, research, and education communication in the event of a military strike taking down the phones.

A quick succession of innovations transformed the ARPANET into something more robust. Two years later, in 1971, came electronic mail, designed by Ray Tomlinson, a pioneering computer scientist on the ARPANET, who wanted to create a means for people on different computers to send messages to each other. Stanford researchers Vinton Cerf, Yogen Dalal, and Carl Sunshine soon established a methodology for sending information over a network, calling it a "transmission control protocol / internet protocol (TCP/IP) network." They had also coined the word that would then be used to describe this new world: the internet.

To connect from one computer to another, Kremen had to make a file transfer protocol, or FTP, request, essentially asking another machine to send him back files. With his position, he could make a request to a German military site, which would reply with a screen for him to log on. Even though he couldn't get in, just the sheer transaction—the knocking on the door of this computer halfway across the world—felt like some weird kind of teleportation made real. It felt as thrilling as connecting with someone around the world on his ham radio. *Oh my God!* he thought, staring at the blinking lights before him, *there's something out there.*

Much of the community, what there was at the time, congregated on Usenet, a free system of discussion boards that had launched in 1980. Kremen surfed through the fifty or sixty Usenet boards, also known as newsgroups, that were up and running, reading them over a long day at his desk: discussions devoted to *Star Trek*, computer games, and sex. It wasn't just discussions, he saw, there were people selling things to each other: albums, clothes, posters. For the hell of it, he bought himself a gun from someone online, just to see if he could. It gave a glimpse, he figured, of a future when people could buy all kinds of things online.

Kremen's mind had a way of racing ahead; perhaps it was undiagnosed attention deficit disorder, but it felt like he couldn't keep internally still. There was so much to do, to create, to innovate, and now, with all this online, it felt staggering. One week during a break, he decided he needed some physical and mental space, and drove off into the desert outside L.A. alone. For five days, he fasted, only drinking water, and reading science fiction novels under the endless blue sky.

The future, he sensed, was exploding around him. This internet, this new frontier, like the endless rippling ribbons of heat rolling over the red desert floor, unfolding. This was still so larval, so new, so lonely in a way, like a tiny new bar in back of a small town. But Kremen saw something in it, a future of community and commerce, a world where people could socialize and shop. There were times in his life when he had little epiphanies, and this was one of them: "money is going to be made from this." He wanted one of those people making money to be him. This meant leaving his job to make this dream a reality. And he knew the best place to make that dream come true: Silicon Valley.

In the fall of 1987, there was one newcomer wandering across the Stanford University campus who was especially eager to stake his claim: Kremen, who was there to begin his master's in business administration. Twenty-five years old, dressed in jeans and a T-shirt, he soaked up the thrill of being there, the smell of the eucalyptus trees and the sight of the immaculate sandstone buildings with red-tile rooftops.

Getting accepted to such a prestigious school, he knew, wouldn't be easy. Stanford's MBA program was among the smallest and most competitive—with only about three hundred students, half that of Harvard or Wharton. So he had taken what he thought was a counterintuitive approach to his application. "People's natural instinct is to tell people how great they are," as he later put it, "but the most successful people understand what someone's problems are and try to solve them, and make a contribution." The problem Stanford had, he surmised, was finding students who weren't going to embarrass the school. So he played up how his unique combination of engineering skills, nonprofit experience, and business savvy would enable him to make the greatest contributions of all. The plan worked, he was accepted.

But as he wandered into the mixer for new MBA students, he couldn't help but feel like a fish out of water again. Despite his best efforts, he was still underdressed. Everyone seemed to be in a suit but him. He wasn't just the worst dressed person there, he noticed, he was the youngest and most socially awkward. As he looked around the room at the people with their shiny suits, firm handshakes, and toothy grins, schmoozing so deftly under the Intel and IBM signs on the wall, he felt his stomach sink as if he was back in Lincolnwood in a room full of Sheldons. *What a bunch of assholes,* he thought, as he headed for the snacks.

It wasn't just the students who rubbed him the wrong way, it was the professors. They seemed stuck in the corporate mind-set, rather than the kind of entrepreneurial spirit he admired. Most didn't have external email accounts, or, to Kremen's astonishment, even know what the internet was at all. Kremen felt like he was from some other generation that just hadn't taken hold yet. Fortunately for him, however, there were like-minded eccentrics among the MBA program. Together, they explored the Bay Area, taking in Dead shows, dropping acid, hanging out at dive bars in Palo Alto and San Francisco.

Even among the oddballs, Kremen was particularly "playful and kooky," as his friend Simon Glinsky put it. He had a voracious, and offbeat, curiosity—questioning possibilities and opportunities at a frenetic pace. At a Giants game, he turned to his college buddy, Phil Van Munching, and asked him if a fly ball that fell into an earthquake fissure on the field would be a ground rule double. Whenever Van Munching owed Kremen money, Kremen would write something embarrassing on the back of the check—like "I don't wear underwear"—so that Munching would have to sign his name above it in order to get the cash.

Glinsky and Kremen joined the Stanford Follies, a musical theater group on campus. Kremen had an idea for a production. There was a metal abstract sculpture in the middle of campus of a group of birds, though to Kremen it looked more like bird shit. The Bird had become a meeting spot, and one day, under Kremen's direction, the Follies showed up dressed like giant birds and camped it up absurdly around the mystified students. He was charmingly uncouth, chastising a friend for taking so long cutting the fat from a steak. "You're being ridiculous," Kremen said, jabbing the fat with a fork and eating it himself.

For competitive brainiacs in the early 1980s, there was one place to go to live the dream: Wall Street, which is just where Gary Kremen found himself in the summer of 1982, at age nineteen. It was the summer after his freshman year at Stanford, and Kremen's high marks had gotten him recruited for an internship at the most renowned Wall Street firm of all, Goldman Sachs, in New York City.

But Kremen, as always, was an outsider from the moment he showed up at 85 Broad Street downtown. He had to ask someone on the subway to show him how to knot his tie. For a guy who'd only dressed up for bar mitzvahs or funerals, reporting for duty in a suit felt like being straitjacketed. The macho fraternity of Reaganites didn't help. Kremen bristled being the lowest pledge on the rung, ordered to pick up someone's laundry or go to an apartment to feed someone's cat. Unwilling to play the game, he'd mumble under his breath—"Oh this is another idiot"—and start showing up increasingly disheveled, hip-hop cranked on his Walkman headphones. After a few weeks, the boss finally pulled him off the trading floor and offered to buy out the rest of Kremen's summer internship, if he'd just go away.

Happy to ditch the job and his suit, Kremen returned prematurely to Palo Alto with what he later called a "capitalist migraine," and an even greater conviction to pursue his true passion: computers. For Kremen, being part of the dawning internet subculture of Silicon Valley had a similarly rebellious feel. The frontier was so new that there was barely anyone really there. The real learning was to be done off campus, so Kremen took consulting jobs writing business plans for start-ups around town, figuring out how to help others raise cash. Burdened with more than $50,000 in loans, he was struggling to get by, and needed whatever income he could muster.

Just after graduation, he founded a business of his own. Kremen was sharing a house in the hills of nearby Los Altos with Ben Dubin, a young engineer who worked for Sun Microsystems, the pioneering computer maker. It was a scrappy house, with a pool and hot tub in the back, and a lock that never had a key. They'd spend their nights getting high in the bubbling water, plotting the businesses they could start. When Dubin told Kremen that Sun was auctioning off old UNIX computers, the two camped out all night until the first sales were available the next morning, and bought several on their credit cards, for $600 each. Dubin then got to work, refurbishing the computers so they could flip them for $2,500 apiece. They advertised the computers in the back of computer hobbyist magazines, and on newsgroups. It was one of the earliest examples of what would later be called e-commerce, except this was as informal as it could get—just a kind of glorified classified ad, using the computer as the medium instead of a newspaper.

But there were enough early adopters to make Kremen's strategy effective, so he got an idea for expansion. People were sharing their own software on Usenet—say, checkbook balancing programs, games, calculators—like a swap meet. Again, Kremen's approach to business was to solve people's problems, and he saw one in all this. People off the internet had a problem because they didn't know how to get the software that was being distributed online. *Why don't I take some of the software on Usenet, put in on tapes, and sell it to people who don't have internet access?* Kremen thought.

Kremen and Dubin called their company Full Source Software. They began downloading the programs online, then bundling them onto an old computer tape, and selling them through ads in computer magazines—forty programs on one tape for $99. The people who wrote the programs weren't getting any of the cash, but this was an unregulated world, and Kremen had rationalized that they were already giving their stuff away for free.

Sure enough, his hunch was right—and he began selling bundle after bundle of UNIX programs by mail. Before long, he and Dubin were making a couple grand a month each, nothing wild, but enough for them to live on. Customers began writing and requesting specific programs to be included, and Kremen expanded their slate to meet their needs and cover their student loans. They started writing their own antivirus programs to include in the bundles, and, as an added incentive, games. The games became so popular that people were ordering the bundles for their companies just so they could get the games for themselves. Finally it hit him: his customers were just a bunch of geeky guys like himself. Geeky guys like games, and he knew there was one other thing they wanted as well: naked pictures, just like the boys of Lincolnwood wanted *Playboys*.

Here was the thing that Kremen understood. If someone really wanted to see naked pictures, all he had to do was buy a porn magazine at the local convenience store. But that required two things: money and, for most guys, some degree of shame. Few people wanted to have to ask some clerk to hand him a copy of *Juggs* from behind the counter. That was a problem waiting to be solved. People wanted to consume sexual content as discreetly as possible. And, in the late 1980s, there was no more discreet way to do this than online.

From the early days of the internet, people were trying in vain to see naked pictures online. Problem was, the bandwidth posed a challenge for viewing images. Image files were simply too big to be uploaded and downloaded given the limited access most early adopters had to the net. At first, people didn't bother with images at all. Instead, they created so-called ASCII art. ASCII stood for American Standard Code for Information Interchange, and basically meant the standard typographic characters on the keyboard: letters, numbers, commas, and so on. When deftly combined the rows of characters could form images—almost like pointillism but with typography instead. It was, in effect, a subcultural predecessor of the emoticons that would later proliferate online.

While ASCII porn still proliferated, surfers had moved on more recently to sharing pages scanned (at a copy shop, since there were few scanners in the homes) from porn magazines—even though it required tremendous patience and effort. Since it would take too long to upload and download a photo image, people distributed the pictures as raw data, or binary files, that could be reassembled using special programs. In practice, compressing the binaries felt like playing some kind of privately dirty puzzle game—and the sheer novelty of getting dirty pictures over a computer provided motivation enough.

Someone would sit at his computer, surf over to a newsgroup such as alt.binaries.pictures.erotica, then see a heading for, say, the latest *Playboy* centerfold. There would then be a series of links to differing binary files, which had to be downloaded one by one, then put back together through software such as UnStuffit. It took several minutes, in all, but the reward felt tantalizing, as the finished image slowly materialized, one row of pixels at a time, until it filled the screen in all its fleshy deliciousness. Or, for people who didn't have the will or skills to do all that, there was Kremen—who, accurately, figured that throwing some naked photos on his compilation tapes would be just the Cracker Jack prize to entice people around the country to order his wares.

That wasn't all customers wanted. They wanted computer security programs. So Kremen started another company, Los Altos Technologies, with Dubin as chief technical officer, devoted to making such wares. Renting an office in a crappy building in Palo Alto, Kremen raised money from an investor to hire a CEO. The CEO had religious beliefs and wasn't thrilled to find out about the porn included on the discs. "We're having a meeting," he told Kremen, "and we're gonna talk about what nobody wants to talk about—this behavior goes against Jesus." But Kremen took it in stride. Having a short attention span created challenges in his personal life, but manifested in his professional life as it did for many like him, as a serial entrepreneur. The thrill, for Kremen, was the inception and creation, coming up with an idea, writing the business plan, bringing it to life, cashing in, then moving on to the next enterprise. And he already had another one in mind.

The idea had come, as they often do, by seeing something new, and realizing what he could do with it. It came one day when he was in his office taking orders for Los Altos Technologies. At first, many of the orders came in over the fax, but they also began taking them via the burgeoning platform of email—still relatively uncommon, given that, in 1992, only about 23 percent of U.S. households had computers. As the emails began coming in, Kremen felt naturally curious about the people on the other end, particularly when he got one from a woman. "Do you think she's cute?" he asked Dubin.

"What does this have to do with anything?" asked Dubin, who was more interested in the size of the transaction. "Look," Dubin said, "this is a $500 purchase order for the software, this is good."

"Yeah," Kremen went on, in his excited nasally voice, "but do you think this one is cute?"

And there was a new way he could find out. The year before, computer scientist Dr. Nathaniel Borenstein, a researcher at Bell Communications, wanted to come up with a way to email pictures of his grandchildren to his family and friends, but this was not yet possible. So he went about creating an email format called the Multipurpose Internet Mail Extensions, or MIME, which allowed someone to attach a file to an email message. Earlier that year, on March 11, 1992, he sent the first one—a photograph of his barbershop quartet, four long-bearded programmers at Bell who called themselves the Telephone Chords, along with an audio clip of them singing their own harmonic tune, "Let Me Send You Email," to the tune of "Let Me Call You Sweetheart."

Innovation often follows a similar path: someone fills a personal need that also meets a universal need. And email attachments would be just that. As Kremen and other pioneers knew, this was a huge deal—the email equivalent of Alexander Graham Bell's first phone call message: "Mr. Watson—come here—I want to see you." *The fucking world is going to change,* Kremen thought. People could do more than just trade pictures of their grandkids, they could trade pictures of each other. So that if he wanted to know what this mysterious woman looked like on the other end of the internet, all she had to do was go to a copy shop, scan a picture of herself on a disc, put it on her computer, and attach it as a file to him. "I wonder if I can convince her to find the closest Kinko's and scan a picture of herself?" he joked to Dubin.

The ideas came in a cascade, one leading to the next, a flash of succession, like firecrackers sparking each other to explode. *What if I can have her send the picture?* Kremen thought, *and what if there are other guys who were interested in her picture?* He could send the photo to them and see if they thought she was cute too. And if he could do that, he realized, *What if I could charge single people to get each other's pictures?*

There was a problem, Kremen realized, that needed to be filled: dating. And he knew this from firsthand experience. Like a lot of busy young people, dating had always been a challenge for Kremen—finding the time, the right person, the magic. Kremen had dated enough to have met and connected with his share of women, but never found his perfect match. Dubin tried to encourage Kremen to change his habits to increase his odds. Stop eating a dinner of just bacon, Dubin would suggest. Don't pretend that swallowing a shot of Scope for breakfast constituted morning hygiene. Oh, and next time you go on a date, you might want to make sure your stained T-shirt isn't inside out.

But Kremen had a different tack. Instead of changing who he was at heart, he wanted to find someone who would accept him. And so, like a lot of people, he tried personals: spending a couple hundred dollars a month on taking out ads in the San Francisco papers, and using the 900 number party lines. He approached it creatively—such as taking out an ad with his friend Simon Glinsky.

Glinsky had become a management consultant, or as he put it, a "corporate therapist," helping companies get along better and make decisions. But, like Kremen, he was also looking for love. Since Glinsky was gay, Kremen had a novel idea of how they might work together to find relationships: by placing an ad for the two of them seeking an opposite pair: a gay guy for Glinsky, and a straight woman for Kremen. Glinsky thought it was a stroke of marketing genius because it would, as he put it, "take the edge off meeting the stranger."

"Double Date, Double Fun," read the headline of their ad in the *San Francisco Bay Times*, "Two friends: gay man looking for gay man, and straight man looking for straight woman, seek fun and supportive double date. . . . Both are successful entrepreneurs; enjoy music, food, hiking, and are in touch with their feelings." Even better, the plan worked. Kremen and Glinsky had several dates—dates that had a built-in buffer that made them all the more easygoing and fun. But no matter how clever Kremen was in his marketing, personal ads had problems of their own. With classifieds, there was no way of knowing what the person actually looked like. Plus, the profiles were so small and limited in scope. With this new technology though, he realized, there could be a way to change all that.

People could also write their own personalized profiles, and include them along with their photos. He could harness the internet for human relationships in a way that no one had done before. He could build a business from a new frontier of dating: online. *This is insane,* he thought, *this is bigger than the vacuum cleaner!* And, of course, if he built it, then he could use it too. As his friend Phil Van Munching put it, "That became part of the mythology, the guy who couldn't get a date invented this love machine." And, Kremen hoped with all his heart, this machine would finally help him meet his match.

CHAPTER 3

THE CON MAN

In the late 1980s, one of the most valuable properties in America and "home to some of the most filthy rich people around," as the *Los Angeles Times* put it, was Cowan Heights, a tony enclave in Southern California. Nestled in the foothills of Tustin about an hour southeast of L.A., the 10,400 residents treasured their sweeping canyon views, their first-class schools, and their exclusive privacy. As local Sharon Thompson, a proud mother of three who'd been living there for fourteen years, told the *Times*, "We're the best kept secret in Orange County."

But there was another secret there too. One summer day in 1998, a grandfather was pushing his grandchild in a stroller along Brier Lane, one of Cowan's main winding roads, when he saw something shocking: a couple having sex behind the window of a sprawling, gray ranch home, in plain view of anyone who happened to pass by. The other windows of the tract home were blackened or covered with aluminum foil. Neighbors began noticing a strange pattern on weekends, as lines of expensive cars began parking out in front of the house, and dozens of couples came and went into the wee hours of the morning. "These are not your run-of-the-mill sleazy people here," Orange County sheriff's investigator Gary Jones later told the *Los Angeles Times,* after receiving complaints about the frequent parties. "We watched Rolls-Royces and Corvettes pull up," he said, "and some very good-looking people."

On February 9, 1989, Jones launched an undercover investigation, dispatching two sheriff's deputies, Charles Daly and Karen Bruner, who, posing as a couple, attended a party at the home one weekend night. After paying an initial $50 membership fee and $40 entrance fee, Daly and Bruner found what appeared to be a makeshift disco inside. The dining room had been converted into a dance floor, with couples boogying under the lights to thumping beats. Homemade porn films were projected on a wall. A short, stocky man in his early forties bragged to Daly and Bruner that the films had been shot right there inside the house.

Daly and Bruner followed the crowd streaming in and out of the garage, which had a wet bar, couches, and tables with pizza and takeout chicken. Many of the women dancing were topless. As they headed toward the bedrooms of the 2,600-square-foot house, they could hear what Daly described as "a lot of moaning and groaning." They followed the thick beige carpeted halls toward rooms lined in aluminum foil and illuminated by red lights. The bedrooms had been divided into smaller cubbyholes, mattresses on the floors, and couples having all variations of sex inside.

As they made their way into the backyard, they found more couples lounging by the pool and in the Jacuzzi, nude. Daly panicked, not wanting to blow his cover. So he did the only thing he could do. He removed his clothes and slipped into the Jacuzzi, where he made idle conversation with the portly man, the host of the swingers parties here at "The Club," as it was called, and the subject of the police investigation: Stephen Michael Cohen.

"My whole life was based on sex," as Cohen would later put it. And though it's hard to believe anything a lifelong con man says, this much about one of the shrewdest hustlers in the history of the internet was true.

Born on February 23, 1948, he grew up in Van Nuys, California, the heart of the porn industry, this self-described "Valley boy" had plenty to fire his imagination. Foremost was his father, David, a flashy Beverly Hills accountant who drove a Rolls-Royce and ran away with his young secretary when Cohen was a boy. He was left with three sisters and an overbearing mother, Renee, who often told him, "You'll never amount to anything." Cohen both despised and admired his father—rebelling against him at every opportunity, while devoting himself to winning the same rewards, though by considerably less scrupulous means.

While other kids in high school were playing ball or hitting the library, Cohen's favorite haunt was a neighborhood strip joint—or, as he put it, a "beaver place"—called the Cue Club. That's where, in the mid-1960s, he got introduced to the sex business. The club owner's son, inspired by the dawning hippie age, had recently started a loose-knit organization called the Los Angeles Free Love Society, with the hopes of a sort of dating service for casual sex. When Cohen learned that the guy hadn't done anything with it, he took it over—placing ads in the *Los Angeles Free Press*, an alternative newspaper distributed on Sunset Avenue.

The $20 membership fee would get buyers a catalog of the Los

Angeles Free Love Society members who were looking for action. Ostensibly, at least. "Between you and I," as Cohen later said, "it was a scam." It didn't take long before Cohen saw that the memberships coming in from the ads were almost entirely from guys. He realized the problem facing any dating service, or club for that matter, was how to get the women.

The answer came to him during a psychology course on the power of manipulation. Cohen listened intently as the teacher began explaining how "you can take words and twist them to where it makes them sound like something, when in reality there was no substance to them," Cohen recalled. "I had a hard time knowing what the fuck this guy was talking about." But it crystallized. "I got the concept," he went on, "you can entice people to take something that's not actually what you're selling."

In other words, he could sell bullshit as long as people were willing to buy it. In this case, he knew guys wanted to get laid, so he needed names and numbers of women. He devised a quick fix: getting the names of the girls at a couple local sororities, and putting them on the list. Cohen had taken out a post office box, and eagerly checked it every day for responses. Day after day he waited, but none came. Perhaps he didn't have a future after all. But one afternoon when he went in to shut down the P.O. box, the mailman handed him a bag of mail—dozens of envelopes with $20 checks inside.

The scam worked well enough for the entrepreneurial high schooler to make his first small fortune, until the inevitable exposure. "All these horny guys started calling them and they didn't know what the fuck it was," he later recalled. The sorority girls' fathers got wise and chased Cohen down—but not before the aspiring sex peddler had made his first small fortune, and completed his first lesson in what would serve him a lifetime: the art of the con.

And no matter how he felt all those years after being abandoned by his father and denigrated by his mother, he found a way, he resolved, to amount to something after all.

Eighteen hundred miles away from Los Angeles, on January 25, 1978, a blizzard hit Chicago that would transform the life of Stephen Cohen and generations to come. Nicknamed the "White Hurricane," the storm paralyzed the region for days: burying homes, killing dozens, and stranding tens of thousands inside their homes, including thirty-year-old IBM programmer Ward Christensen. So he decided to spend his snow day working on a side project for his club, the Chicago Area Computer Hobbyists' Exchange, or CACHE.

Christensen had recently imagined a novel way to replace the bulletin board at their club. Instead of tacking 3x5 cards on the cork board with messages like "need ride to next meeting" or "let's get a group purchase of memory chips," what if they had a computerized way of doing that instead. Innovation isn't just about invention. It's about vision—understanding the available tools and seeing a way they can be used together to make something new.

At the time, the idea of an online community was like science fiction—it did not yet exist. These were still the larval days of what would become the internet. Computers were largely refrigerator-sized mainframes. But in 1978, there were two recent innovations on Christensen's mind: the Hayes internal modem, an early hobbyist device for exchanging data between computers, and the S-100 microcomputer, a personal machine that preceded the introduction of IBM and Apple PCs. The challenge was to bring these tools together in what he described as a "Computerized Bulletin Board System," or CBBS. And with the White Hurricane, he had time on his hands to finally do it, so he called up CACHE buddy Randy Suess to help with the hardware.

Working through the storm, they had it up and running in two weeks. Sitting at a computer, someone would simply dial up a phone number through her modem, which would connect her to the CBBS machine. Once there, she could leave her own text messages, which would get displayed on other hobbyists' home computers when they dialed up. There were no photos or videos or audio files, just text. Though the system could support only one caller at a time, it achieved Christensen's original mission: bringing the old idea of a cork board and thumbtacks into the dawning digital era.

After several months of testing among the CACHE crew, they introduced it to a national audience in an article they cowrote in the November 1978 issue of *Byte*, a hobbyist magazine. "The Computerized Hobbyist Bulletin Board System," they wrote, "is a personal computer based system for message communication among experimenters. People with terminals or computers equipped with modems call in to leave and retrieve messages." It marked nothing less than the birth of the online community. With a modem, a microcomputer, and some phone lines, anyone could create a BBS. For the first time the technical means of online community had been created and described for the masses, or, at least, the early adopting "experimenters." And one of the first was a thirty-year-old con artist in California who saw just the way to exploit this new technology, with sex. And his name was Stephen Cohen.

By the time he heard about the invention of the BBS in the fall of 1978, Cohen was in need of a fresh start. Since his first con with the Los Angeles Free Love Society, he had spent the next decade in broken relationships and breaking the law. He was already on his third wife, Susan Boydston, after two short, failed marriages that left him with bitter ex-wives and three estranged kids. Some of what he claimed to be doing at the time seemed too outrageous to believe. If asked, Cohen would claim he'd spent time studying law at (unaccredited) schools such as the Southern California School of Law, and Western State. He claimed that he also had traveled to Panama, where he'd taken the bar and become admitted as an attorney. He insisted, even more, that he had several meetings with Panamanian military leader Manuel Noriega, offering legal advice.

One thing, however, was certain, he was running afoul of the authorities. There were a string of cases: he got convicted of petty theft, forgery, pled guilty to writing bad checks, and served a year in state prison. And yet, with his old persuasive skills, he always managed to win over the judges enough that their hearts went out to him. "I think it is a terrible shame that a boy like you finds yourself in this predicament," a judge told him during a hearing in 1977, when he was convicted of grand theft and forgery. "I don't know anything about your background," the judge went on, "but I strongly suspect that you come from a good family. . . . I think you can make it and be an awfully nice citizen if you will just work at it."

Alas, Cohen didn't. The chip on his shoulder was too big to give in anymore to anyone else's idea of who or what he should become. He would prove himself according to his own terms. He would fulfill his own sense of empowerment, his belief that he alone was smarter than everyone else, able to outwit them, out-invent them, and out-hustle them. So just as soon as he could get to it, he looked for his next scheme.

Since he was a kid, Cohen had always been something of a geek, tinkering around with gadgets and taking a couple of computer programming courses at local colleges. He was among the early phone "phreaks," hobbyists who circumvented the telecommunications system to get free long-distance calling, among other things. Though he wasn't, by even his own accounts, much of a programmer, he had a hacker mind-set in that he could see where and how to exploit new systems to his advantage. And so, when he read about this new innovation of Bulletin Board Systems, the thirty-one-year-old had just the idea for how to cash in: by launching a kind of Free Love Society for the dawning computer age in the summer of 1979. He called it the French Connection BBS.

While most, if not all, of the few dozen BBSs at the time were free services devoted to the conversation of early adopting hobbyists—computer nerds, Grateful Dead fans, sci-fi geeks—Cohen had a distinctly for-profit model. He would charge $18 a month for subscriptions, which would give members access to leave messages for each other on a variety of sexual and special interest discussion boards. Log on and press 6 for "Sexual Discussions," 9 for "Guys' Locker Room," S for "Swing Scene," and so on. The BBS didn't discriminate between straight, gay, or anywhere else on the spectrum. As long as a customer paid for the chance to get laid, that was fine by him.

Cohen created the BBS, with the help of some freelancer programmers, and ran it from a computer in a bedroom inside the home he'd recently bought in a gated community of Orange County. To promote the site, he relied on word-of-mouth, posting messages on competing BBSs, such as the popular Los Angeles service West Side. Once again, he faced the same challenge as he had with the Free Love Society—getting women to entice the men. So he offered free membership to women and, more expediently, he logged on under assumed female names. "He would sit there and type in different names and be different people," Boydston later recalled, "always women. He was trying to attract men customers."

He would secretly pose as "Tammy," one of the system operators, or sysop, the title for someone who ran a BBS, or even

Boydston herself, leaving messages for horny guys, and dealing with customer service. He began taking computer programming classes at a local college, where he gave speeches about his BBS, as much to show his knowledge as to recruit new members, including his professor.

Slowly but surely, the French Connection acquired members and curiosity seekers. "Tammy" would flirt with them and take their credit card numbers, or checks, which Cohen would have sent to a variety of post office boxes. When he needed a name for the company that owned French Connection, he made a nod to his mother, as if this were all about proving her wrong. YNATA, he and a friend decided to call it, as in "You'll Never Amount to Anything." As Cohen later explained, "We were going to build something that would prove to the world we weren't losers."

Despite his outlaw ways, Cohen amounted to something historically significant as a result of the French Connection. He had not only pioneered online dating, he had figured out something that would have huge significance in the decades to come: how to get people to spend money online. The answer was simple, as he told his wife, when she asked why everything he did seemed to have to do with sex. "Sex sells," he said.

But that wasn't all he was selling in the early 1980s. While growing the French Connection, he was pulling off a garden variety of small-time schemes. He opened a liquor store in Santa Ana, created a computer timeshare business. He became a repo man, starting a car repossession company, cutting new keys on machines he kept in his garage, then flipping the cars for profit. He got a private eye license, so he could get intel on enemies, created an answering service, a limo company, and travel agency (which he used to hustle free and discount flights for himself and his friends). When Boydston started inquiring about his frequent international trips, Cohen told her, with a straight face, that he was working for the CIA.

Though he burned through businesses and money, often stiffing the people he owed, Cohen always managed to keep up appearances—determined to show to others, if not himself, that he did amount to someone significant, just as his father did. Using Boydston's money, he got a two-story home near the golf course of Coto de Caza, an upscale, five-thousand-acre community in the Saddleback Mountains in south Orange County. He even leased the same car his dad drove in Beverly Hills: a Rolls-Royce.

It wasn't just the riches he craved, it was the lifestyle. "He wanted to be Hugh Hefner," as Boydston later recalled. And, unbeknownst to her, he began living his own version of the *Playboy* lifestyle. After peopling the French Connection with fake women, including himself, he ended up drawing enough real women to the service that they began holding their own regular meetups in person—gathering for pizza, drinks, and the inevitable more. "I'm sitting at home wondering what could be taking this man's time so much," Boydston recalled. "Finally I find out he's been swinging."

By 1985, Boydston finally had enough. She had grown to despise the man, considered him short-tempered, easily frustrated, and always looking for a new con game. She felt as if he thought everyone, including her, was less intelligent than him. And when she deigned to confront him, she said, he could be threatening. She considered him a sociopath, unable to process feelings, guilt, or remorse. "He doesn't know what love is," she later said, "he only knows what's working for him." She divorced him.

Cohen was who he was by now. And his perspective on marriage was either cynical or liberated, depending on one's point of view. "You got to understand," as he later said, "in this world where we live in, there are a lot of people that go into relationships, and most relationships don't last. Most marriages don't last." And that's where he saw a lifestyle, and a business. "People kind of lose sight of how to get back into the game," as he put it. "I offered an opportunity where they could do that."

The French Connection was just the start. The opportunity he gave the wealthy residents of Orange County was their very own swingers club. From the moment he opened The Club over the July 4th holiday weekend of 1998, it felt like the fantasy world he'd always wanted had come to life. He had spread the word clandestinely, but effectively, as talk of the jet-set sex parties made its way around the golf courses, health clubs, and PTA meetings in town.

Prospective members had to meet a specially appointed liaison at a grocery store downhill from the house on Brier Lane, where a trusted representative would screen them, then escort them to the house with blackened windows. There, Stephen Cohen, the ebullient host, dressed in his terrycloth robe and a fat cigar in his mouth, warmly welcomed them to the party. Cohen took pains to lasciviously decorate the home as a suburban sex palace: the dance floor, the cubbyhole bedrooms, the red-lit halls. There were buckets of Kentucky Fried Chicken at the ready, fresh towels by the Jacuzzi, and satellite TV showing sports and porn.

By early 1989, he had gone from ten couples on a Saturday night to as many as sixty. The growing demand had to be met. Volunteers doing loads of wash, arranging the catering, getting the music, even adding air ducts to the bedrooms to improve the flow, even The Club's own baby-sitting service, for $15 a night, including a meal. With more people came more rules, which Cohen posted in his monthly newsletter: no drinking, eating, or smoking in "swing rooms," no nonbiodegradable lubricants, no answering the front door, no sex for husbands after their wives have fallen asleep.

There were monthly themed events: Movie Night ("You get to make that special, private movie with that special person. . . . Please bring an empty VHS"), Sensuous Massage Party, Lace and Lust Lingerie Party, and Dirty Dancing ("Two ladies will do their thing on the dance floor with that special pole"). And for those members with an interest in computers, Cohen offered complimentary memberships to the French Connection. "We are the place where people go," Cohen promised, "and everyone cums!"

But among those people coming, he learned, were undercover police. After sheriff's deputies Daly and Bruner secretly attended three parties at The Club, the Orange County team moved in against Cohen, citing him for safety violations—such as the partitioned bedrooms—and then, in May 1991, hitting him with misdemeanor public nuisance. "They acknowledge what they're doing, and they're not ashamed," as one angered resident told a reporter at the time. "They're rather brazen in their openness."

Cohen, characteristically, pushed back, arguing that his was a private club, not a business, and therefore the county had no right to legislate them for being swingers. "The sex isn't the issue," said Orange County Deputy District Attorney Kimberly Menninger. "This is a very residential area. If you had a tennis club, you'd have to get a permit."

But, of course, sex very much was the issue, and suddenly the man who'd fashioned himself as the Hugh Hefner of BBS swingers was challenging the community to look at their own standards. "Does anyone think it's wrong to have sex outside of marriage?" Cohen's attorney, William J. Kopeny, who had represented high-profile murderers, rapists, and, soon, the cops accused of beating Rodney King, asked residents during jury selection in the Orange

County courthouse. One potential juror said she couldn't be objective, because her "husband was a cheater." Another asked to be relieved because he was a schoolteacher. And yet for some, the subject of swinging was not a concern. "I am a Christian. I don't do it myself," as one woman said, but "I'm not a judge of other people's choices."

During the trial, Cohen likened The Club to the Boy Scouts, a private organization and not a business that violated zoning. Cohen fashioned himself as a freedom fighter, standing up for his "alternative lifestyle" that was being wrongly discriminated against by the county. "This trial has nothing to do with sex," he told reporters outside the courthouse. "These were lawful activities behind closed doors." His attorney went on, "They have a lot of moral outrage but very little evidence" that it was a business. After two days of deliberation, the jury had enough people who believed this that they became deadlocked. With a mistrial declared, the charges were dropped. "I think it was proper," Cohen told the reporters. The county had "put on a lot of interesting evidence," he went on, cockily, "but none of it had to do with the law."

But, in fact, there was little victory for Cohen, whose life was spiraling out of control. The scams and schemes were finally catching up with him. The owner of the house he was renting terminated his lease, leaving his swinging days over. Microsoft, Lotus, and other companies sued him for copyright infringement for distributing pirated copies of their software on the French Connection. He was under investigation for bankruptcy fraud, falsifying documents and loans.

He'd even been caught impersonating an attorney. It had happened when, one day in L.A., he walked into the federal courthouse, carrying a briefcase, his clients in tow, as he represented them in front of the judge. In the midst of all this, his father, David, had died of a massive heart attack—the same thing that had killed Cohen's grandfather. Though the two had become estranged, the loss of his father left him with a permanent sense of failure for never having made his father proud. "My father was a very decent person," as Cohen later put it, "unfortunately I didn't turn out that way."

By the fall of 1991, Cohen had gone from Hugh Hefner to Howard Hughes, a recluse who barely left home, or changed his clothes. The previous year, he'd gotten married at a swingers convention in Vegas to Karon Poer, a single mother he'd met at The Club. At first, she admired him. She thought he was a lawyer with a nice home. He took her to Panama. He told her he had to get someone to sign a document attesting that he was an attorney in that country. She waited in the car while he went into a hospital with his computer, sweltering in the heat. Finally he came back, and told her, with a laugh, that the doctor he was supposed to meet wasn't there, and no one else spoke English, so he couldn't get his form completed. Why he needed some doctor to authenticate him as a Panamanian attorney was beyond her.

But then she began noticing unusual behavior. He spent most of his time in his home office. She'd bring him a sandwich, the only time she was allowed in, and find him in his swivel chair, papers strewn around the floor, a small round table holding a copying machine. He ordered a bunch of large cell phones and a strange machine in the mail, and she watched him take it apart. Then he drove with her to a parking lot by a freeway. "We're going to be leaving in a few minutes," he told her. "I'm just using this machine." As cars passed, numbers appeared on the machine, which he wrote down.

He returned home, and she saw him programming the cell phones with the numbers. Then people would come over and buy them for $500. Poer asked one of the women who came by what this was about, and she told her Cohen had programmed the phones with other people's numbers so anyone who bought them could make free calls. Poer thought she wasn't smart enough to understand what they were doing, but knew it had to be wrong. She felt like Cohen was wasting himself on bad things. "He's so smart and could make millions," she later said. "Why did he always want to do something wrong?"

In less than a year, their marriage was falling apart. Poer barely saw her husband, who spent his days and nights locked in an upstairs bedroom, with his computer, from which he ran the French Connection BBS, and his multiple phone lines. She'd overhear him answering the phone and pretending to be someone else: Steve Johnson one day, Frank Butler the next. When she asked why, as she later recalled, he'd snap "That's my business!" Paranoid and prone to outbursts, Cohen had outfitted their home with security cameras, and refused to let her answer the phone or the front door. "Steve was a lonely man," recalled Poer.

Finally she woke one day to find the house surrounded by police. Cohen told her not to be scared. As the doorbell rang, he snuck out the back, and tried to get to his car to escape—but to no avail. Cohen's run had come to an end. On October 21, 1991, as part of his previous stints impersonating attorneys and attempting to bilk an elderly woman out of $200,000, Cohen was convicted of bankruptcy fraud, making false statements, and obstruction of justice. During the trial, Special Assistant U.S. Attorney Elizabeth Hartwig admonished Cohen for his cocky assurance that he could always wriggle free. "He still thinks he can flimflam and flummox his way out of this," she said. "He thinks he's the one that can go into any court and outsmart anybody." But not this time. Cohen was sentenced to forty-six months in prison.

After a stay at a prison in San Diego, Cohen soon joined the other inmates behind the barbed wire fence inside the gray and red brick penitentiary in Lompoc, California, about an hour north of Santa Barbara. As he sat in his cell, he longed for the days of The Club—the sex, the "sucking and fucking," as he liked to put it. But there was still something out there at least, something that was his: the French Connection, which was still running online. And he wasn't about to give that up for anyone.

One day, Poer answered the phone to find Cohen on the other line. He wanted to take care of her, he said, give her the money he owed so that she and her son could live. I'll give you the French Connection, he told her, you can make money from it. "I have no idea how to run that," she told him, incredulously.

"I'll call you and talk you through it," he assured her.

He had a friend set up the phone lines in her new apartment, and the calls immediately started ringing day and night from angry creditors. They'd ask for Cohen, or one of his many aliases—Steve Johnson, or Tammy. When she told them Cohen was away, they pressed her for the information they could never get from him: Where does he live? What kind of life does he lead? Where does he keep the Rolls-Royce? Cohen told Poer what to say to each person, whom to answer, whom to ignore, whom to say they had the wrong number. People who had bought the cell phones with the fake numbers were complaining that they were no longer working.

When she wasn't fielding calls from them, she was trudging back and forth to the Lake Forest office where the French Connection was running. He told her how to press a button on the computer to see how many people were online. No one used their real names on the BBS. One person was just called Hot2. Poer's name on there was 900Babe. Cohen had always told people there were five hundred online at a given time but she saw maybe a few dozen at best. She claimed he told her half the people online at a given moment were fake, to give the appearance of a crowd. He had her offer a lifetime special membership to try to raise extra cash. Despite everything Cohen had always boasted about the French Connection, it was barely making enough money for her to buy groceries, she later claimed.

Cohen was working in the prison library, helping with the computer systems, which gave him access to log into the French Connection. Cohen would call Poer and tell her when a computer was acting up, and she'd trudge down to the office at Lake Forest. He'd keep her on the line as he told her how to clean the computer disc to get the system up and running again. But Poer was quickly losing patience. There were too many lines to manage, too many phone bills to pay, it was wasting her time, her money, trying to keep this going with her kid in school. But Cohen had planned for this too—and left a password at the phone company, so that no one but him, certainly not Poer, could turn off the lines without his permission.

When she tried calling Cohen back one night in prison, however, the administrator at Lompoc said there was no way Cohen had been calling her. Prisoners were only allowed five minutes on the phone, she was told, there's no way it could have been him. "I know his voice," Poer replied, "he was on the phone." When the prison checked on Cohen, however, they found him in bed, where he insisted he had been the whole time.

Poer stormed over to Lompoc soon after, and told Cohen she was done tending to the French Connection and his other matters. "If you're not changing your ways, there's no more me and you," she said. Cohen's arrogance hadn't left him, despite him sitting there in his prison orange. He laughed, ignoring her. "When they came to my room to check on me when you called," he told her, "I told them I was sleeping." He asked her for quarters that he could use for the vending machines.

As Poer would later claim in court, he continued to somehow hack the phone system in prison to call her beyond his allotted time, filling her answering machine and pager. She claimed she was in a tanning booth one day when her pager rang twice with a West Virginia area code. Concerned her family was having a problem, she called back only to find it was Cohen—who'd somehow spoofed the number from which he was dialing. It wasn't until she turned over her phone records that they believed her. Running a business out of prison was a violation, and Cohen was threatened with solitary confinement.

Finally one day she got calls from angry French Connection members who said the service was down. Poer drove out to the office, only to find the computers were gone. She figured Cohen had arranged for someone to steal them to cover for the service's failure. In fact, Cohen accused Poer of trying to take over the business, running him out and putting it in her name. So he dispatched a trio of associates to get into the office, and get his computers back. "I authorized the repossession from incarceration," as he later put it during a deposition, and moved to the offices of a friend. When asked if she was relieved it was over, later by a court, she replied "I couldn't care less." She was filing for divorce, her time with Cohen was done. She said she could never trust anyone again now that she had been through the ringer with him.

Many who knew Cohen described him as a loner. Bonnie Hite, an acquaintance of Cohen's who bought a used computer from him in the mid-1980s, recalled the many times he offered to help her with her system—not as a way of making a move on her, but simply because it seemed like he wanted companionship. "Steve was kind of the lonely soul," she would recall, "he had no place to go really." With every day behind bars, his golden years running The Club seemed to fade. He was just a forty-three-year-old man, divorced again, with nothing to his name—nothing, that is, except the French Connection.

As he waited for his release in late 1994, he lay in bed at night imagining how he would bring it back to life, and reclaim his throne. Three and a half years is a long time behind bars, but it's even longer in the fast-moving world of technology, he knew, and he feared being left behind. Sometimes he would hustle some phone time to call people on the outside and find out what was new.

Among those he called was Steve Grande. Grande was an engineer who took the more traditional path, with graduate studies in computer science at Boston University. In the early 1980s, he owned a company that sold timeshares on mainframe computers, and had bought software from Cohen to run on his machines, and they had remained friends. Like Cohen, he was among the early pioneers online, and ran a large BBS called Liberty—initially for fellow libertarians to talk politics, but later dedicated to their greater interest: computer games.

When he found out that Cohen was running the French Connection, he blanched at the sexual content, but found common ground in shop talk: technical issues, marketing, billing. Grande kept Cohen at arm's length when he'd start bragging about his sexual conquests or his Rolls-Royce. But whenever the French Connection ran into technical problems, Cohen would call him for advice. Cohen seemed like a real gearhead to Grande, a guy who always had random computer and phone equipment in his garage when needed. Grande was surprised when, on one occasion, he drove out to Cohen's house to get some parts, and found this otherwise disheveled guy chomping on a cigar and living in a mansion in a gated community. Cohen seemed to love any nitty-gritty technical thing, Grande marveled. They both shared an interest in locks, and spent days taking mail order lock-picking courses, tearing apart locks and putting them back together.

Like Hite, Grande and his wife, Barbara Cepinko, took pity on Cohen. They let him store his computer in the office of the temp agency Cepinko owned, Midcom, for the time being. Grande took his occasional calls from prison. There was something about staying up on the new advances that kept Cohen engaged, and encouraged. The world was always moving ahead, and he wanted to be right there with it the moment he was free. "So," he would ask Grande, "what's going on in the computer world?"

CHAPTER 4

MATCHED

For Kremen, the key to success in business was about serving an essential human need. In his career, he would invest in more than fifty companies specifically according to the Hierarchy of Needs described by psychologist Abraham Maslow in his seminal 1943 paper on motivational theory.

To find happiness, Maslow wrote, a person must meet five primary needs, which he represents as levels of a pyramid. First come physiological needs: food, water, warmth, sex. Next is safety: security of property, health, employment, and so on. The third is love/belonging, such as friendship, family, and intimacy. Then there's esteem—confidence, achievement, respect—and, finally, self-actualization, in the form of morality, creativity, spontaneity, and, at last, acceptance. Building businesses that fulfilled these needs for others was the key to success, Kremen believed. "The closer you are to Maslow's Hierarchy of Needs," he said, "the better you do."

That's what Kremen thought one day in business school. Love and sex were right up there on the hierarchy of needs, and, even better, they were big business too. It was early 1993, and Kremen, rumpled and red-eyed, was leafing through a towering stack of periodicals inside the Stanford Business Library. He was researching his latest epiphany: that he could use this new technology, specifically email and email attachments, to build a new kind of business online.

There was no time to waste. Kremen could feel the pressure of his impending thirtieth birthday upon him. Being in Silicon Valley during this time didn't make it any easier. He was part of a small group in the community buzzing over the release of Mosaic, software that radically transformed how the internet could be navigated. Developed at the University of Illinois Urbana-Champaign, Mosaic was among the first browsing software for what was being called the world wide web. Once downloaded, it allowed someone to view the web as a series of interconnected pages with links from one to the other. For Kremen and other early adopters, it signaled the dawn of a friendlier, potentially more mainstream online experience.

The competition to exploit the burgeoning new world online was everywhere, the sleek MBAs angling to be the masters of the computer age. He had to make something of his life, his education; prove himself to his mother, Harriett, show her that he could amount to something, become the success he was always destined to be. And, of course, maybe he could meet the love of his life too—a woman who would do what sometimes it felt like his mother never did: love him unconditionally. He knew that he was no different than anyone. Everyone wanted love and connection. All it took was someone who cared enough to bring it to them.

He could be that person. This could be just the ticket. The market for dating, he learned, was already huge, and growing. More than fifty million people in the country were single, and, as the coveted twenty-five- to forty-nine-year-old demographic, they spent plenty of cash, an estimated $3 billion, on the two primary means of finding each other: personal ads and dating services. And, Kremen noted, it was new technology that made this possible in the first place. It had happened in the 1980s with the advent of voicemail, which allowed people to leave messages for each other when they answered personal ads in newspapers. There were nearly two thousand publications offering personal ads, which, at around $120 a pop, were contributing as much as 40 percent of the publications' total revenues.

It wasn't just the newspapers that were making cash from love connections. The 900 numbers, the pay-per-call party lines that began with the 900 prefix, were charging upward of $2 per minute for people who could leave and respond to messages for each other. They were also seeking out dating services, matchmaking companies such as Great Expectations. Customers would go into facilities and record a video of themselves, describing themselves and their ideal partners, and then search through the existing videos—at an annual membership cost of $2,000. Kremen also found what he called "Fiddler on the Roof" matchmakers, professional love gurus—at a price of $10,000 per match.

But, as Kremen knew from his own experience with personal ads and 900 numbers, the business of love was seriously flawed. The newspaper ads, limited to around thirty words, gave people little room to express themselves, let alone any way to share photos and voice messages. He thought about all the times he'd answered ads expecting to meet, say, the statuesque blonde described in the paper—only to find anything but, in reality. Even worse, searching through the ads was time-consuming and inefficient. The dating services were just as frustrating, requiring someone to physically go into an office, and subject him- or herself to the embarrassment of telling one's heart's desires to a stranger.

As he walked across the Stanford campus, past the coeds coming and going to class, he thought about how the computer revolution was going to bring them together in ways they couldn't conceive. His key insight was this: the internet of the future wasn't just about connecting with information, its real power was in connecting people with each other. He could use a computer to store all the information of the daters, building detailed profiles based on how they filled out questionnaires. The ads could stay there in the database for as long as needed, allowing someone to search, anywhere, anytime, for a potential match. Billing and processing could all be done privately, with anonymous email addresses maintaining discretion. With real photos, there'd be no more guessing if someone really looked like Halle Berry or Brad Pitt. And if someone found what they liked, they didn't have to wait for days or weeks to hear a response, they could get matched in seconds.

And, Kremen realized, online matchmaking could just be the start of an empire. Classified ads, including personals, represented a whopping 40 percent of revenues for the newspaper business. He could take the entire model of classified ads and move them to the nascent internet. Personals would just be the start, then he could create an online service for everything and anything someone might be seeking: jobs, apartments, motorcycles, anything. *Holy crap,* he thought, stopping in his tracks. This dating service could be the proof-of-concept for online classifieds. Electronic Matchmaker would be its name.

But there was one big problem, he was broke. With $50,000 in debt for student loans, there was every reason for him to just take the easy route. Classmates from Stanford were out working at big firms such as McKinsey and First Boston, moving into giant homes in Palo Alto, and driving their Porsches and Ferraris. But, just as he wasn't cut out for Goldman Sachs, he knew he wasn't going to be happy with that life either. He didn't care whether he fit in or not, he never really did after all. But he wanted fulfillment, happiness, love, like anyone. He was willing to risk everything to try to make it happen. Kremen sold his stock in Los Altos Technologies, pocketing $25,000, and took out $2,500 on his credit card. But to make this dream a reality, he'd have to convince others that his idea wasn't crazy, and to join his crusade to build the love machine.

Peng Ong, an industrious artificial intelligence researcher from Cambridge, Massachusetts, was attending a seminar for young entrepreneurs at Stanford, when he saw someone stand out from the crowd: a wild-eyed, wild-haired portly young guy with a nasally Chicago accent, Gary Kremen. Kremen was in a group talking about his experience at Los Altos Technologies and struck Ong as just the kind of outlandish visionary he might learn something from that day.

He was right. As Kremen spoke rapidly about his plans for an online classified company, Ong perked up at the idea of, as he put it, "technology meets relationships." With his experience in programming, he could provide Kremen the technical side of the start-up. He struck a deal to become Kremen's cofounder and chief technical officer of the company they now called Electric Classifieds, taking 10 percent of the ownership. Living in an apartment in a drug-infested part of San Francisco on Cole Street, Kremen subsisted on two burritos a day as he began bringing Electric Classifieds to life.

The first thing he had to do was raise money, which meant writing a business proposal. "ROMANCE—LOVE—SEX—MARRIAGE AND RELATIONSHIPS," he typed at his PC in the summer of 1994, "American business has long understood that people knock the doors down for dignified and effective services that fulfill these most powerful human needs. People everywhere have met partners at bars, at sports fields, at work, from friends and through personal classified advertising. Electric Classifieds, Inc. (ECI) has identified a significant opportunity to fulfill these needs more effectively and rapidly using the 'Information Superhighway,'" he wrote, referring to the nickname of the internet at the time.

With more and more people joining nascent online services, such as America Online, CompuServe, and Prodigy, and the demand for dating, he projected that Electric Classifieds would become a $50 million company in less than five years. To make it happen, he would need $2 million, and an initial investment of $300,000 to develop and roll out Electronic Matchmaker, the dating service on ECI.

Making stacks of copies of his proposal at a nearby Kinko's, he sent the business plan around town. He'd call his friends from Stanford, one by one, and ask, "You know any rich people who make investments?" There were plenty of skeptics. Even around the Stanford community, the potential of online dating—let alone online commerce at all—didn't register. Some would give Kremen a name, and he'd follow up as best he could, calling the switchboard of the company, asking for the person. "Hey, my name is Gary Kremen," he'd say, "we should talk about this really cool thing I'm working on."

Many of the recipients weren't even using email yet, and simply tossed his proposal in the trash. But there was one important person who found it intriguing: Ron Posner. A Harvard MBA with a math degree, Posner had already established himself as one of Silicon Valley's most prescient angel investors and CEOs, having led high-profile consolidations in the computer security business. To

Posner, Kremen wasn't crazy at all; he was a visionary who stood out for his indefatigable belief in himself and his ideas. He thought Kremen's idea for online dating was, as he later put it, "just off the wall enough to succeed." Kremen helped stoke his interest by suggesting, with some exaggeration, that other big guns in town were intrigued too. Posner was in—for $50,000.

The investment from Posner was just the "social proof," as Kremen put it, that Kremen needed to get others interested. He'd go around town pitching the company, and dropping Posner's name. If someone else showed even a glimmer of interest, Kremen would monopolize that too, dropping that person's name to other potential investors. And so on. It was a process of "ham and egging," as he called it, piling on names to establish credibility. When need be, he appealed to a potential investor's personal interest in meeting that special someone. "Hey," he'd say, "I'll get you a date!"

Kremen had no interest in dressing up for his pitches. He'd just wake up, grab clothes from a pile in the corner, maybe eat a bit of toothpaste from a tube to freshen up, then show up to meet with guys in Armani suits. He'd make the rounds in his stained T-shirt and ratty jeans. No matter, with Posner's vote of confidence, people were taking a more active interest in him, and in what he was creating. He found himself getting wined and dined by potential investors. By early 1994, Kremen had raised $200,000—which got the attention of others who wanted a piece of the action.

One night, he was being courted by a potential investor over dinner at the home of Vinod Khosla, cofounder of Sun Microsystems. "This is one of the best business plans I've ever seen," Kremen was told. But there was just one problem: Kremen's inexperience. There was still an old boys' club in Silicon Valley, and the money guys wanted one of their own in control. "Gary," the investor said, "we like you, we're interested in funding this, in fact, we want to fund this—but we want you to have a professional CEO, someone who has done this before in a big way."

Kremen wasn't so sure he wanted any suits in his company, but agreed to go out to an airfield in Palo Alto to meet a potential CEO. Kremen wondered why the airfield, when he saw a small prop plane—and its pilot, the CEO. But something seemed amiss. Half of the man's body was paralyzed from a recent stroke. Kremen wasn't a big flier, and this just made his anxiety even worse as he climbed into the passenger seat while the CEO took the controls.

As they flew over the Valley, the CEO carried on about his accomplishments, bragging in a way that rubbed Kremen wrong. Between the guy's arrogance and the erratic flying, he walked away from the experience with an even greater resolve to do his own thing—even if that meant turning away a fortune. "Fuck this shit," Kremen said to himself. "I'm gonna be the boss." His instinct proved right, when he found a smaller investment firm, Canaan Partners, that was willing not only to let him be his own CEO—but invest $1.7 million. Kremen was in the money.

With the seven-figure investment, Kremen needed a place to set up shop. At the time, the internet start-up culture in San Francisco seemed almost nonexistent. So when Kremen looked for an office, he felt like he had become a pioneer for real. And if no one else was there with him yet, so be it, he knew it was only a matter of time before others joined in. He found a dingy basement office, with windows just poking up over the sidewalk, in a burgeoning section of San Francisco known as South Park, an area south of Market Street that surrounded the namesake park, known for drug dealers. The only other businesses Kremen saw in the area were

Chinese tailors, and a printing company. He worked around the clock, eating cold burritos and sleeping on the floor.

One day he and the engineer he'd hired, Kevin Kunzelman, who had recently dropped out of the PhD computer science program at Cornell, were hanging out in the basement offices, smoking a joint and cranking the Dead, when Kremen had a vision: the Old West. "In 1804, where would you go first? Think about it," he said, pausing wide-eyed to take a hit, filling his lungs with the smoke, choking back a cough, then exhaling like the dragon he felt he was riding at the moment. "Do you go up to the hills?" he went on. "A lot of people went up to them, the mountains, start logging, and that turned out to be worth very much. That's where the money was, but maybe the smart people got that land at the bottom of the Golden Gate Bridge. There are pioneers, and there are settlers. I'm going to be the settler."

On the budding internet, the land was represented by domain names: the online equivalent of a physical address. Each domain name—a string of characters such as home.txt—represented a destination on the internet, say, a specific computer or a website—a system that dated back more than a decade to the days of the U.S. government's ARPANET. At first, to keep track of the domain names, SRI International, a nonprofit research institute established by Stanford University, created the Network Information Center, or InterNIC. But in 1991, with about twelve thousand domain names registered, the U.S. Defense Information Systems Agency awarded the contract to a Herndon, Virginia–based technology company called Network Solutions, Inc., or NSI. By 1993, NSI, under a grant from the National Science Foundation, was the sole registrar for domain names online.

Getting a .com, .org, or .net address was just a matter of filling out a form and sending in a request and, best of all, it was free.

For Kremen, a domain meant having a place to own, to build businesses upon, to sell. It didn't matter if he was building on there yet, he had to get the domains, now. In 1994, it was a "land grab," as he put it, and the domains were the land. He couldn't help but laugh to himself. Having been early on the net, he knew that, as he put it, "no one was thinking about commercial shit." And if they were, they were afraid to bust a move.

Grabbing a copy of the *San Francisco Bay Guardian*, he flipped to the classifieds in the back and ran his finger along the categories: autos, housing, jobs. His plan was simple, to register each and every category of classified ads online. He didn't know what he'd do with the domains, maybe build a business, maybe sell them. He'd figure that out later. For now, he went on a mission to register as many as he could. One by one, he emailed the registration forms to Network Solutions to lock up the domains—and when he hit the limit of how many he could register under his own name, he hit up his friends. One day, Phil Van Munching got a call from his old friend Kremen. "Dude, you have to grab some domains," Kremen told him.

"I don't have money to buy houses," Van Munching replied, "what are you talking about?"

"No, no, no," Kremen told him, "the internet!"

Van Munching had been around Kremen long enough to know that he could come off like the mad genius, and he knew better than to misread him. Kremen saw two steps ahead of everyone else, and he trusted him implicitly. He would never forget how, hours before his wedding, Kremen took his hand, and held it for a minute, reassuringly. As Van Munching later said, "I thought it was the strangest, the most wonderful thing that had ever happened to me from a friend." So when Kremen suggested he register domains, he figured he was onto something. "Grab some domains, do what I'm doing," Kremen told him. "This is the future!"

Within weeks, Kremen had registered more than two dozen under his own name including Jobs.com, Housing.com, Autos .com, and the one he chose for his online match-making company, Match.com. And, also, as an afterthought, he registered one more domain, inspired by the adult ads in the back of the *Bay Guardian*. On May 9, 1994, he registered Sex.com by sending in an email and a certified letter to NSI. He didn't know what, if anything, he'd ever do with it. But for now, he had a bigger focus: launching the first dating site online, Match.com.

"Anal sex... Abstinence... Animal rights... Very conservative... Marijuana OK... Children should be given guidelines... Totally open and honest even at the cost of being thought rude . . . Religion guides my life . . . Make charitable contributions . . . Not happy with job . . . Night thinker . . . Would initiate hugs if I wasn't so shy . . . Can meet most of my own needs . . . Enjoy a good argument . . . Have to-do lists that seldom get done . . . Sweet food, baked goods . . . Sports in stadium . . . Artificial or missing limbs . . . Over 300 pounds . . . Drag . . . Exploring my orientation . . . Women should pay . . . To be happy, I must love my match."

There was another four-letter word for love, Kremen knew, and it was data, the stuff he would use to match people. No one had done this, so he had to start from scratch, drawing on instinct and his own experience with dating. Generating data—based on the interests of a person in categories such as the ones he was typing out on his PC ("Mice/gerbils or similar . . . Smooth torso/ not hairy body")—would be the key to the success of Match; it was what would distinguish electronic dating from all other forms. They could gather data about each client—attributes, interests, desires for mates—and then compare them with other clients to make matches. With a computer and the internet, they could eliminate the inefficiencies of thousands of years of analog dating, the shyness, the missed cues, the posturing. They would provide customers with a questionnaire, generate a series of answers, then pair up daters based on how well their preferences aligned.

While Peng Ong and Kevin Kunzelman developed programming in the fall of 1994, Kremen worked late into the night in his office creating the questionnaire that each Match customer would fill out in order to generate the necessary data. He started from his own experience—putting down the questions that mattered to him: education, style of humor, occupation, and so on. With the help of others, the headings on the list grew—religious identity/observance, behavior/thinking—along with the subcategories, fourteen alone under the heading of "Active Role in Political/Social Movements" ("Free international trade . . . gender equality . . ."). Before long, there were more than seventy-five categories of questions, including one devoted to sex—down to the most specific of interests ("Muscle domination presupposes a substantial difference of physical strength between the two lovers, and a mutual interest in using this as a part of erotic games or lovemaking").

But then the more he thought about it, he came to an important realization: he wasn't the customer. In fact, no guys were the customers. While men would be writing the checks for the service, they wouldn't be doing anything if there weren't women there. Women, then, were his true targets, because, as he put it, "every woman would bring a hundred geeky guys." Therefore, his goal was clear, but incredibly daunting: he had to make a dating service that was friendly to women, who represented just about 10 percent of those online at the time. According to the latest stats at the time, the typical computer user was unmarried and at a computer forty hours a week, so the opportunity seemed ripe.

Not only that, online dating could explode thanks to a new and considerably more accessible way of navigating the internet. In December 1994, Netscape Communications, an outgrowth of the Mosaic team, released a powerful new web browser called Navigator. Navigator transformed the budding web into an even more fluid and alluring experience, with faster-loading pages of graphics and text that created a familiar experience, almost like reading through a magazine. A company called Yahoo!, founded by two Stanford grad students, was created to help people search for the page they wanted. Kremen knew immediately that these were game changers—and ones that, as soon as possible, he would implement with Match—which, for the time being, would be designed as an email service.

To enrich his research into what women would want in such an innovation, he sought out women's input himself, asking everyone he knew—friends, family, even stopping women on the street—what kinds of qualities they were looking for in a match. It was an essential moment, letting go of his own ego, understanding that the best way to build his market was to enlist people who knew more than him: women.

In his mind, if he could just put himself in their shoes, he could figure out their problems, and give them what they needed. He'd hand over his questionnaire, eager to get their input—only to see them scrunch up their faces and say "ewwww." The explicit sexual questions went down with a thud, and the notion that they would use their real names—and photos—seemed clueless. Many didn't want some random guys to see their pictures online along with their real names, let alone suffer the embarrassment of family and friends. "I don't want anyone to know my real name," they'd say, "what if my dad saw it?"

Kremen went back to Ong and Kevin, and had them implement privacy features that would mask a customer's real email address for an anonymous one on the service. But there was a bigger problem: he needed a female perspective on his team. He reached out to Fran Maeir, a former classmate from Stanford business school. Maeir, a brash, dark-haired mother of two, had always been compelled, albeit warily, by Kremen—"his fanaticism, his energy, his intensity, his competition," as she put it. When he ran into her at a Stanford event and told her about his new venture, he was just as revved. "We're bringing classifieds onto the internet," he told her, and explained that he wanted her to do "gender-based marketing" for Match.

Maeir, who'd been working at Clorox and AAA, jumped at the chance to get in on the new world online as the director of marketing. To her, Kremen's passion and pioneering spirit felt infectious. And the fact that he was turning over the reins to her felt refreshingly empowering, given the boys' club she had been used to in business. Maeir showed up to the basement office with pizza and Chinese food and got to work. One day, an engineer at Match asked her "what weight categories do you want in the questionnaire?" She arched her brow. "Oh no," she said, "we're not asking that." Women never want to put down their weight, she explained to the dubious guys. Instead, she had them include a category for body type—athletic, slim, tall, and so on. She also cut down Kremen's intimidating laundry list of questions. Fewer questions enticed more people to register, which meant a larger database, and a greater selection of potential matches.

But they had a catch-22. Women weren't going to join unless there were other women online. Maeir, along with other women brought on to help spread the word, started with their friends. They created a friendly logo—a radiant red heart inside a purple circle—and printed up promotional brochures. To entice people to try out the service, they held promotional events at happy hours in Palo Alto, only to have, as Match marketing executive Alexandra Bailliere put it, "thirty guys with pocket protectors and no women in sight." Trish McDermott, a marketing executive who'd worked for an upscale matchmaking firm and had founded a dating business trade association, and the others would slip on fake wedding bands to ward off the guys. "Are you interested in meeting new people?" she'd say. "This is a new dating site, like personals in the newspaper but it's on the internet." Then she'd get a blank stare as the person would ask, "What's the internet?"

They weren't just targeting heterosexual women, they were going for the lesbian, gay, bisexual, and transgender community. Match's marketing consultant Simon Glinsky pointed out to Kremen how his gay community had already been early adopters online, using bulletin boards and the nascent communities such as America Online, CompuServe, and Prodigy for dating. Glinsky related from his own experience, having grown up in Georgia, where meeting other gays was a struggle.

Glinsky went to a gay computer club, where members gathered to talk about AOL, and the latest deals at Radio Shack. It was in a small theater south of Market Street. He sat onstage with their computers as they explained Match to the crowd. "Now we're debuting this new service," Glinsky explained, "it's going to be online. You could match; you could make good use of your time to meet the right people. You could eventually meet, we've a lot of people on there and you could see a lot. I think I remember saying it would be free but there would be subscriptions later."

They held a promotion during a gay skate night at a roller rink in Burlingame, just north of Palo Alto. Bailliere and Glinsky urged skaters to come over and learn more about Match, offering to take their photos with giant digital cameras—which seemed exotic at the time. One by one, the skaters marveled at seeing their faces appear on the computers, and word, in fact, began to spread. The *San Francisco Examiner* ran an early piece on Match, speculating this could transform the "grand old dating game," as it put it. "What happens when singles have an alternative to bars," the article went on, "and don't just meet based on first impression/physical attractiveness alone?"

On April 21, 1995, Kremen launched Match.com. He had taken out a patent for another one of his formidable inventions: dynamic web pages, the means through which information could be entered and processed online. Match was a free service, supported by ads, with the idea to charge for subscriptions when it grew. And there was only one way for it to get there. "We need more women!" as Kremen shouted, storming through their basement office. "Everyone wants to go to a party where there's women!" he said. "Every woman means ten guys join!"

Since they didn't have real women, they had to create some themselves. Maeir dispatched interns to the Usenet groups, where they posted laudatory reviews of Match. When *Rolling Stone* wanted to run a piece on Match, along with a sample profile of a female member, the women at the office scrambled to invent one. Bailliere drew the short straw, slipped a black jacket over a white T-shirt and smiled for the camera. Her fake profile, "Sally," said she was seeking a twenty-five- to thirty-five-year-old guy for an Activities Partner, Short Term Romance, or Long Term Romance to "go hiking and have LOTS of fun."

Having her profile, albeit fake, in a high-profile magazine sent a stream of messages to the email she'd set up. A German in Brazil told her he wanted to use her to re-create Nazi youth camps, and became so obsessive that she became nervous. "Gary," she told Kremen, "I don't know who this person is or if he's really even in Brazil." Concerned, they worked with consultants to develop safety guidelines, such as meeting prospective men from the internet in public places. Maeir had them market Match as "safe, anonymous, and fun." They also invented self-policing tools for people on Match—such as giving them the ability to block and report others for bad behavior.

The site's PR executive, McDermott, began hosting a weekly chat session called "Tuesdays with Trish" to dole out dating advice. She billed Match as the dating solution for the emerging online generation. "We're delaying marriage," she'd tell reporters. "Many of us moved away from home, and many were just moving from suburbs and starting careers and we lost all that fabric of informal match matching when we stay home. . . . You can put a profile up this morning and that night have a response waiting for you."

It wasn't just Match they were marketing, it was their lovelorn leader, Kremen—who, at every opportunity, advanced Match as the fix for his own romantic quest. "I started the company because I decided it was the best way, maybe the only way, to find the best woman in the world," he told the *Mercury News*. "It was a fantasy thing, and I was tired of eating and drinking alone." Kremen's love life became a running narrative. He was the Charlie Brown of the internet. "Will Gary Kremen's success . . . help him find a date for Saturday night?" read a headline in *Websight Magazine*. With all the business, "Kremen may not find his perfect woman any time soon," as *Interactive Week* put it. Or, as Kremen told *San Francisco Focus*,

"I'm a success-oriented person. It kind of irks me that I haven't met someone yet."

Managing Kremen for the media proved a challenge, however, for McDermott. She implored him to wear certain colors that looked good on camera, no patterns or stripes, only to see him show up in a dirty tie-dye T-shirt. She felt overwhelmed by him, how smart he was, a mind that moved a mile a minute, with an almost avant-garde sense of direction and purpose. When she watched him tell one reporter that he was going to bring more love to the world than Jesus, she gulped. But there was something about him that made her trust what he was saying: his authenticity, she realized, was a lot more interesting than some executive who had been coached.

The press ate it up. There were profiles of Kremen and his company in *Wired, Forbes,* the *San Jose Mercury News,* and elsewhere. "If the poet Elizabeth Browning were around today," read a story in the *New York Post,* "do you think she would say to Robert Browning, 'How do I love thee? Let me switch on my Power Mac, send you an E-mail and count the ways.' It's hard to say, but in 1995 the search for romance and flirtation is clearly flourishing on the World Wide Web."

There were soon competitors in the wake—with names like the Singles Online Network and the World Wide Web Dating Game, but Match.com, the pioneer, was dominating. Within a couple months, they'd reached nearly a million hits on their website, with a large chunk of the traffic coming from America Online. They were going back and forth to Kinko's, scanning photos of prospective daters—and tossing the occasional nude shots in the trash. Before long, word spread around the office that two of the people who had met on Match were getting engaged. The group gathered to watch it unfold in a chat room, as the guy made vows straight out of *Men Are from Mars, Women Are from Venus* ("I will not go into my man cave when I need to communicate . . ."). They had to pass a Kleenex box around the office for everyone who was crying.

Kremen was riding higher than ever before. And he wasn't the only one. A new gold rush had begun. It had started, it seemed, on August 9, 1995, with the initial public offering of Netscape Communications, makers of the first mainstream web browser, Navigator. Though the start-up had yet to show a profit, that didn't stop the stampede of investors—who more than doubled the price of the shares on the very first day they were available. Over the course of one day, the company's cofounder, Jim Clark, became worth more than a half billion dollars. And the dream of the new dot-com multimillionaire was born.

By that month, Match had hit more than 10,000 users—and was getting hundreds more each day. Once they crossed the 25,000 threshold, the plan was to start charging for memberships. The plan, as Kremen told *Interactive Week*, was to build out his other domains—such as jobs.com, autos.com, and housing.com—in similar ways, and fulfill his vision of being the classified ads king online. As for his own love life, Kremen claimed that building his empire was consuming all his energy. "Now I'm so busy," he said, "I don't have time to use the service." Privately, however, he still longed for someone with whom he could share his life. But it was easier to distract himself with his business than lament that he had still not found a deep connection of his own.

In the meantime, he focused on his plan. As part of it, he was keeping tabs on his domains, checking up on them now and then to make sure everything was in order. But one day in mid-October 1995, his assistant told him that something unusual had come up when checking on one particular domain that Kremen owned, Sex.com. It seemed like it was actually registered under someone else's name. Kremen hadn't thought much about that site since he'd registered it the year before, but he figured there just must be some mistake.

At the time, anyone could go on a site called WHOIS—as in who is—and type in the name of a website to see the name of the registrant. So Kremen typed "Sex.com" in the empty field on the website, and hit enter. Sure enough, his assistant was right. The name on the registration page for Sex.com was no longer his. As he sat there staring at his screen, mystified, he had one thought: *Who the fuck is Stephen Michael Cohen?*

CHAPTER 5

SCREWED

When Cohen left federal custody in August 1994, after serving three and one half years, he didn't have much: a pair of shoes with holes, a borrowed shirt, old pants that hung loose around his prison-thinned waist, and a bus token out of town. Most of his property was with his ex-wives. He had spent so much of his time behind bars stewing in anger at Poer for leaving him—the old familiar feelings of unworthiness, betrayal, of loved ones treating him like he had never amounted to anything—but now he wanted to start anew. "I sat down with myself and I realized that being enraged had to end," as he later recalled. So he decided to rebuild his life with the only two things he ever needed, a computer and a dream.

It was good timing. The internet had undergone a radical transformation during his time behind bars with the advent of the world wide web. Gone were the hobbyist days of BBSs that he had known. The web was exploding online life into the mainstream. Like Kremen, Cohen had a knack for seeing the business that could come with the technology, and he saw it all as clear as day. In the privacy of their own homes, people could shut the door, turn on a computer, and surf for anything their hearts, and bodies, desired. He knew exactly what they'd be seeking, sex. His vision was the dark side of Kremen's. It was the opposite of love, nothing touchy and feely like online dating, no nonsense about finding true love, it was about lust and release, pleasure in disconnection. And if anyone was going to cash in, it was going to be him.

His plan was to bring the French Connection to the web. While he was in prison, he had left his computers at Midcom, the temp agency owned by his friend Steve Grande's wife, Barbara Cepinko. Now that Cohen was out, however, Grande had no interest in having a sex site run from Midcom. So Cohen picked up his machines, and found the cheapest place he could get: a run-down office in the back of an auto salvage yard. Sitting there among the rusting carcasses of old rides, he set up his computers, pressed power, and heard the comforting rush of the hard drives spin. But when he logged on to the French Connection BBS, the members had long since left, since no one was running it while he was away. All he had left was a ghost town.

Needing cash, Cohen spent the next several months doing consulting work for Midcom: helping with Cepinko's collections, fixing her phone system, and keeping her technology up and running. When the vending machine stopped getting stocked, Cohen volunteered to pick the locks, stock them with snacks on his own, and collect the money. Grande and his wife found Cohen to be an odd guy, someone who showed up dressed in ratty jeans and T-shirts, but who maintained an unusual air of sophistication and snobbery. Cohen told them he was a lawyer, and offered to help sort out legal matters. When talking with problem clients, Cohen would introduce himself on the phone as an attorney and then, in a very professional manner, inform uncooperative callers that he would not hesitate to take legal action against them on Midcom's behalf. If he was talking with lawyers, he would adopt a chummy tone, this was all business, and deftly get the other attorneys on his side.

When he wasn't playing fix-it man or lawyer, he was picking Grande's brain for anything he could learn about the web. Grande had already migrated to the web, and had started TrainWeb, a destination dedicated to "preserving passenger rail heritage." Special interest sites were among the earliest on the web, fueled on the passion of niche communities—such as gamers and sports fans. TrainWeb was for railroad nerds—with pictures of trains and tracks, reviews of rail lines, and links. And Grande was, to Cohen's fascination, making money through a new form of advertising online: banner ads, small strips of advertising that, when clicked, would lead to the advertiser's site.

In October 1994, *HotWired,* one of the first online magazines, began selling the first banner ads, and they became an essential part of the emerging economy. When Cohen saw Grande's banner ads, taken out by other train enthusiast sites, he saw dollar signs. To him, the conclusion was obvious. If someone could make money selling ads to a bunch of locomotive nerds, imagine how much money he could make from porn. What he needed, he realized, was a powerful destination, a website that would attract every horny guy with a computer and an internet connection. And if people wanted to find sex online, he knew, there was one place they'd probably type in first: Sex.com.

Sitting at his PC in the junkyard one October day in 1995, Cohen surfed over to Sex.com but was surprised to find that there was nothing there: no porn, no ads, just a blank page. What moron would own a site as valuable as Sex.com and do nothing with it? He found out one day when he searched the domain name registration database, and typed in a WHOIS search for Sex.com. *Who the fuck,* he wondered, *is Gary Kremen?*

Shortly after, Grande was sitting in his office, working on his TrainWeb site, when Cohen bounded in full energy. "I acquired a domain name!" Cohen told him, excitedly. "Sex.com!" Though Grande had no interest in hearing about Cohen's latest sexcapades, he knew enough to wonder why anyone would let go of a domain as clearly valuable as Sex.com. Cohen didn't go into the details of how he got the site. He just said that the previous owner had simply "lost interest," and was eager to dump it. Whether that was true, Grande had no idea. This was Cohen, after all. The guy was always getting into weird deals, he figured as he watched Cohen go—anything was possible.

WHOIS Stephen Cohen? That's what Kremen wanted to know after he found the guy's name registered to Sex.com that October. There must be some kind of mistake, he figured. But the nascent business and culture of domain names was still unsettled. The past year had seen some contentious battles among the early adopters snatching up domains—but with the rules vague, it remained a Wild West. In 1993, MTV VJ Adam Curry registered MTV.com, informing the company he was staking ground for their future online only to be told, as he later put it, the company had "no interest in the internet." But the company apparently gained interest by October 1994, when it sued Curry, who'd since quit, for the name. Curry, however, was staking his ground on principle, calling the case the "*Roe v. Wade* of the internet."

Curry eventually settled out of court for an undisclosed amount, giving up the domain, but his point was made. MTV weren't the only ones missing the boat. At the time, less than one third of Fortune 500 companies had registered their names online. For a May 1994 article in *Wired*, writer Josh Quittner registered McDonalds.com to prove his point. When he called the company to inform them, a media spokesperson asked, "Are you finding that the internet is a big thing?" Quittner sold them back the domain for $3,500, which he donated to a public school computer fund.

Kremen didn't know why or how Cohen had obtained what was rightfully his. Taking a break from the chaos of Match.com one day, he dialed the number on the domain registration page that Cohen had filled out, and was connected with a secretary who announced the name of the company, Midcom. "May I speak with Stephen Cohen?" Kremen asked.

Cohen was meeting with Cepinko in her office when the secretary buzzed, saying he had a call. Cohen and Cepinko both knew it could be anyone. An ex-wife, an ex-con, some angry creditor, a lawyer, who was it this time? "He says his name is Gary Kremen," the secretary replied.

Cohen always knew when to put on his lawyerly airs, the way he'd threaten Midcom customers with lawsuits if they didn't play ball. So certainly now, with Kremen, there was no reason to pull any punches. He listened smugly to this excitable guy with the nasally Midwestern accent blabber on about how and why Sex.com was under his name, like a poker player with a royal flush and no hurry to show his hand. "I'm a trademark attorney," Cohen told Kremen, "let me help you understand." Kremen believed him, and why wouldn't he? He had no idea that Cohen had spent years impersonating lawyers and falsifying documents, had served three years in prison, and was now on probation for his conviction. He just knew the guy sounded convincing.

Kremen listened intently as Cohen pedantically explained why, in fact, he was wrong. Since 1979, Cohen explained, he had operated a BBS called the French Connection. According to Cohen, French Connection had all kinds of discussion boards devoted to different areas of sexual interest: swinging, kink, and so on. But for those who wanted a general interest hot chat, well then, they went to the discussion board devoted to all kinds of sexual communications—or, as it was nicknamed on the BBS for decades—Sex.com.

To which, Kremen thought: *No way*. There was no dot-com nomenclature in 1979, let alone 1989. This guy was full of shit. But when Kremen balked, Cohen told him that, because of his Sex.com discussion board on the French Connection, NSI, the domain name registration company, duly transferred the Sex.com domain back to its rightful owner, him. Kremen told Cohen to send him the documents from NSI that proved this, but he wasn't backing down. "I'm Sex.com!" Cohen snapped, "you're not Sex.com!"

And then, like a conductor playing the orchestra, like the lawyer playing the chump, Cohen performed for Cepinko like he always did, showing her who was boss, and how this loser on the other end of the line was the one who would never amount to anything. "Go fuck yourself!" Cohen told Kremen, and slammed down the phone.

Go fuck myself? Kremen thought, as he sat at his desk, dumbfounded, listening to the dial tone buzz. Who did this guy think he was? He was so smug, so cocky, so dumb but self-assured. He thought he was better than Kremen. Kremen wanted answers, so he called the one place that would have them: Network Solutions, NSI.

The internet seemed like such an ephemeral thing, all the wires and data, the ones and zeroes, but there was a physicality to it. A business. Building. Bricks. Hallways. People. And across the country in a nondescript office building in Herndon, Virginia, a suburb forty minutes northwest of Washington, was one of the most important ones of all: NSI. The company, which had recently been acquired by Science Applications International Corporation or SAIC, the large, government technology contractor, was still the one stop in charge of the increasingly valuable domain name registration business. And a successful one at that. The company had begun charging $5 per domain name registration in September 1995, and was on its way to making $20 million in the first six months alone.

The company had a policy that conflicts over domains were to be handled between the disputing domains, but Kremen wasn't settling for that answer. After speaking with an agent from NSI, he was transferred to Sherry Prohel in NSI's Business Affairs Office, who assured him she would look into the matter of Stephen Cohen and Sex.com. A few days later, Kremen received a call back from Bob Johnson, who identified himself as the vice president of NSI and SAIC. *Finally*, Kremen thought, *someone who would get to the bottom of this.* Johnson was familiar with the issue at hand, and told Kremen, in fact, that Ms. Prohel, whom Kremen had spoken with the other day, had requested a change order to transfer Sex.com back to him. Kremen felt a wash of relief at the words.

However, Johnson went on, he had overridden her order. "Why would you do that?" Kremen wanted to know. There was a shuffling of papers. Because, Johnson explained, apparently someone at Kremen's former company, Online Classifieds, had authorized Sex.com to be transferred to Cohen—and even accepted money in exchange. *That was absurd!* Kremen thought. *Online Classifieds?* That wasn't even really a company, he knew. That was just the name he had used back in 1994 when he was first registering domain names out of his old apartment. Who at the "company" of Online Classifieds was Johnson talking about? Johnson couldn't say, nor would he grant Kremen's wish of being transferred to a supervisor. That, Johnson demurred, was that. The call was done.

And again, Kremen thought, *what the fuck?* He reached out for the paper on his messy desk with NSI's number, and stabbed the buttons as he dialed right back. "May I speak with Sherry Prohel please?" he demanded. "It's Gary Kremen."

Some clicks, and then Prohel. "Who's Bob Johnson," Kremen asked, "and what's his position at NSI?"

A pause of confusion, the moment swelling. "There is no Bob Johnson working at NSI," she replied. Kremen put two and two together. Bob Johnson was Stephen Cohen.

Sixty miles into the desert southwest of Las Vegas, Sheri's Ranch, one of the many legal brothels in and around Pahrump, Nevada, baked in the sun. For the urgent day trippers from the Strip, Sheri's had the necessary attractions—ten or twenty prostitutes, depending on the season, lounging scantily clad and ready inside the air-conditioning. Starting in the 1970s, the owner, James "Jimmie" Miltenberger, transformed it into his own empire, successful enough to buy him a private plane, yacht, and mansion.

But one day in the fall of 1995, he found himself talking with a short, stocky, smooth-talking entrepreneur with graying hair who had bigger plans for the brothel, Stephen Cohen. Cohen's eye widened as he painted the picture of what he wanted to build, as he put it, "the first-ever sport/adult fantasy resort in America." It would be high-end, costing $100 million to build, and catering to the jet-set visitors in Vegas.

Cohen gestured to the hundreds of acres of undeveloped desert surrounding Miltenberger's dusty acreage. Just picture it: eighteen holes of golf, skeet-shooting, a race track, tennis, squash, convention facilities, and corporate timeshares. Convention center too. But that wasn't all, Cohen went on, building to the good part. Inside the resort would be what Cohen said would be a sexual paradise. He wanted it to be like the recent movie *Exit to Eden*, a comedy starring Dan Aykroyd and Rosie O'Donnell about two undercover cops on a tropical island for swingers and bondage enthusiasts—except this fantasy island would be a short drive from Vegas. Polynesian-themed, with waterfalls and Jacuzzis, it would provide more than five hundred exotic female escorts at the ready to fulfill every imaginable desire—and, for the right price, the ones they couldn't imagine too. And its onomatopoeic name said it all: "Wanaleiya."

Miltenberger, a fixture in and around Vegas recognizable for his smoky sunglasses and jet-black helmet of Wayne Newton hair, had heard a lot of pitches in his time. But this one intrigued him—especially along with the $7 million that Cohen was offering to buy Sheri's Ranch. On October 27, with an agreed upon $25,000 down and $50,000 monthly payments to follow, Miltenberger struck the deal with Cohen. It "sounded pretty good," as Miltenberger later recalled. "They said, 'We want to buy.' I'm going to listen."

By the end of 1995, Cohen was back in action again. It had been weeks since the call from Kremen demanding back the Sex .com domain, a blip on his radar that was already fading to black. He had bigger plans on his mind: building a new empire of sex for the dawning digital age. The French Connection and The Club were just the prequel to what he was hatching now: Sex.com and Wanaleiya.

He explained the plan one day in Anaheim to the board members of his new corporation, Sporting House Management. The gathering included a real estate agent friend, James Powell, who was already involved, as well as Cepinko and Powell's ex-girlfriend, Bonnie Hite, the woman who'd once bought computers from Cohen and now ran Orange County Power Wash, a home cleaning service. Hite and Cepinko had no interest in the sex business, but Cohen, as always, was an able and confident pitchman. "I saw great opportunities in the brothel business, large sources of revenues," as Cohen later recalled. The Mustang Ranch, Nevada's most notorious brothel, was doing more than $25 million annually—$5 million from merchandising alone, "great cash," as Cohen put it. Hite was intrigued—but still skeptical, as she took notes, this was prostitution after all, right?

When Cohen heard doubters, he had the habit of grinning at their naïveté, and then condescendingly schooled them in what he knew. When he spoke of Wanaleiya, he called it "the island." He spoke smoothly. "It may not necessarily be the cup of tea of the people sitting in this room," as he described later during a court case, "but I feel today that there is a need in our society for this type of an entity. The island, I think, is spectacular. I think it would be sold out on a daily basis. A lot of people would pay." Think of it this way, he went on, "a lot of people want to take vacations, men and women that are engaged in business, that have careers, that are not necessarily looking for long-term relationships, have sexual desires, want to go to a place, a place something like a place of Eden where they—it's clean, medically, a place that they can engage in wild and crazy sex that's controlled. For an all-inclusive fee."

He compared it to Hedonism II, the notoriously sexual resort in Jamaica. "It is a hotel that—the average person is naked," he said. "Swinging activity takes place. It's not uncommon to find people engaging in intercourse in the Jacuzzis and on the beach and various places. People come there that are single and meet other singles or married people." Wanaleiya, however, would "be more inclusive. Wanaleiya was going to have a section for married people, strictly for married people, an area for swingers, an area for gays, an area for lesbians. It was going to be a five-star-type resort, something with the class that you would find at the Ritz Hotel to—from the food you ate, to your existence at this place."

Hite was beginning to warm to the idea the more Cohen explained it. Plus, as Hite put it, having two women on the board would bode well for the brothel and, as she later said, "make sure it was all kept very clean." There was the matter, of course, of raising the $7 million. Where was Cohen planning to get that from? Again, he had all the answers. He wanted to take the brothel public. That's right, he went on, the Wanaleiya IPO.

Cohen handed out glossy brochures he'd already prepared: which had the glistening torso of a bikini model on the front. There were charts explaining the financials. California investors with at least $100,000 in income or a net worth of $500,000 could buy in for a minimum of $2,000. The proceeds from the $1 million penny stock offering would go toward marketing, architectural work, securities underwritings, and, he estimated, $85,000 for a "resort theme product line," including "quality fine art reproductions and small collectibles." Another public offering would follow to raise the rest of the $7 million needed to buy Sheri's Ranch.

But that wasn't all, Cohen went on. There was Sex.com. Though he hadn't done anything yet with the domain, the trademark rights alone, he told them, would be worth millions. The internet gold rush, he was convinced, would be fueled by the same thing that had driven humanity for eons: the desire for sex. The trademark assets of Sex.com, he went on, could raise the innate value of the IPO—and allow them to make the fantasy island real. So he proposed a deal: he would transfer the ownership of the Sex.com domain to Sporting House Management in exchange for $2 million down the line. With everyone agreed, Cohen proceeded with the plan—and arranged for a future board meeting at a nearby Olive Garden restaurant.

On January 30, 1996, Cohen announced the IPO in a press release touting how Wanaleiya would "accommodate, at 100% capacity, 500 female companions for 300 patrons. The Company plans to offer full guest access to the resort and its amenities for a single fee of $7,000.00 per weekend." The company issued shares on a certificate illustrated wth a nude bathing Greek goddess, with the name Wanaleiya written in script underneath, along with "the ultimate resort." Plans were already in motion to bring the concept to a real island in the sea. "At present, the Company is negotiating the acquisition of an entire island in the Caribbean for the purpose of developing the Ultimate Fantasy Island," the press release went on. "Imagine, being able to buy your own condominium on Fantasy Island."

"Gaming industry visionaries faced cognitive dissonance resistance when they first arrived in Las Vegas in the 1950s, we may well encounter the same," Cepinko stated in the release. "The American woman has come a long way. In recent times she has freed herself from antiquated, social prejudices and is more self-sufficient in determining her choices. Most of the women we meet express a positive view of the resort. As a matter of interest, we have received an extraordinary number of requests to develop a future adult fantasy resort just for women. We are taking up the concept in earnest."

The prospectus went online at Cohen's new site, www.sex.com. "Purchase stock in a REAL WHOREHOUSE!" it read, alongside photos of naked women and the flashing promise of the phrase: "You will have sex soon!" The others on the board took comfort that Cohen, the man they believed to be an experienced lawyer, included warnings to investors that, despite the bullishness, this was no sure thing. "The securities offered hereby involve a high degree of risk," one read, and another: "Purchase of these securities should be considered only by those persons who can afford to sustain a total loss of their investment." But Cohen's lavish descriptions of a private airstrip, twenty-four waterfall lagoons, and twelve manicured gardens couldn't mask what he was offering: an international coterie of prostitutes. "Even in a bull market," cautioned *USA Today*, in an item about the project, "it will be a tough sell."

Despite the small investments from the board's family and friends, skepticism grew. George Flint, a lobbyist and spokesman for the Nevada Brothel Owners Association, told the *Orange County Register* that their plan "would never work. It doesn't make any sense," he said. "It's too far from Las Vegas. You don't mix conventioneers and the brothel business. They were going to import women from foreign countries—that means green cards and immigration—to be prostitutes?"

Before long, Miltenberger grew doubtful too. Every time he asked for the $7 million, Cohen seemed to have an excuse. "I'd say, 'Cohen, where's the money?' and he'd say, 'next week,' he was going on vacation," Miltenberger told a reporter for the *Register* one afternoon while on his yacht outside San Diego. "I'd say, 'Cohen, where's the money?' and he'd say he was sick and going in the hospital." The last straw came when he went out for dinner and drinks with Cohen—only to have the wannabe sex mogul stiff him on the check. By summer, Miltenberger pulled the deal. "They were supposed to come up with the money. They never did," he said, "so we canceled their option. We told them unequivocally 'no.' Sheri's Ranch is not for sale to them at any price. I want nothing to do with them."

Cohen's dream of a fantasy island was dead. When Cepinko was later asked by a lawyer during a subsequent court case, at what time did she realize her stock in Sporting House was worthless, she replied, "The day he handed it to me."

"Really?" The lawyer demurred. "Did you figure it was some vast smoke-blowing activity?"

"It was Steve Cohen," she said.

Once again, Cohen found himself empty-handed. No empire. No board. No millions. But he did have something. When the IPO and the corporation fell apart, so did the deal for Sporting House to buy the Sex.com domain for $2 million. The coveted site was still his. For all his faults, Cohen had one admirable quality: his resilience. He'd found new wives, new ventures, new innovations. He had already invented online dating, he told himself, imagine what he could do now, on this great new frontier of the web, with Sex.com. And no one was going to stop him.

While Cohen was finding himself on his own by early 1996, so was the other guy who called himself the inventor of online dating, Gary Kremen.

Since his initial, ill-fated run-in with Steve Cohen over Sex .com, which he decided to deal with when he had the time, all of his attention was consumed by running Match. The site was barreling toward 100,000 members—enough to soon start charging for subscriptions at a rate of $7.95 per month. With $7.5 million in a second round of venture and corporate capital from Canaan and others, the company took over another floor of their office building in South Park, which was now attracting more and more dot-coms hoping to become the next Netscape.

But Kremen relished that, this time, he was in the lead. More and more companies were launching online—Amazon, for books, Craigslist, for classifieds, eBay, for used goods. Sponsors were lining up on Match to help drive up subscriptions—Joe Boxer, Godiva, Clinique, and others. Deals were struck to distribute Match through major media sites, including *Women's Wear Daily* and the *New York Daily News*. For Kremen, it felt like he was finally living the dream he'd been angling for since his days selling *Playboys* and Hubba Bubba in Lincolnwood.

With a salary of $80,000, a lot to him at the time, he could finally start paying off his student loans, and socking away cash. And he was even finding love himself. He'd met a girl, offline, whom he'd been dating. She was feisty, a feminist who balked at wearing lipstick because, as she told him, "it's like putting a vagina on your mouth, and it's a symbol of male oppression." Getting approval from his mother, Harriett, though, remained elusive. When he showed her a story about him in *Forbes* magazine, she said, "What about *The Wall Street Journal*?"

She wasn't the only one giving him a hard time. So was the Match.com board. Though Kremen was the chair, he was feeling besieged. He saw this as a media company, they saw it as a technology company. They wanted to sell their software to newspapers, who could use it for their own classifieds. But Kremen felt this was a losing plan. At first, he played along, taking meetings with major newspaper publishers, who didn't even seem to understand the internet. Kremen would show them screenshots of Match, since they weren't often online—but they just scrunched their faces in confusion. He returned to his board, disgusted and dismissive. "They're a slow-moving dinosaur," he said. "They're road kill." The board just looked at him like he was a punk kid. "Oh, they've been around for a hundred years," he'd hear in return. "They'll be around for another hundred years."

Kremen thought some of the board members were painfully out-of-touch if not homophobic. Knowing around 10 percent of Match members were coming from the gay and lesbian community, Kremen wanted to expand into that market—but the board balked. He tried to appeal to their greed. "Look, let's talk about business here," he'd say. "This is an underserved market." But they wouldn't budge. This only cemented Kremen's disdain over having to answer to others. "Guys with gold make the rules," he'd say.

Torn between his own visions and the realities of the board, the stress started getting to him. He began lashing out at employees, shouting, insisting on going over and over the slightest details of the site. He'd tacked a flowchart on his wall, each page of Match spread out before him, as he edited it over and over again. He had no patience for office politics, the endless complaints of one employee about another. "God, listen to their fucking problems," he'd say to himself. Or, to them, "You're wasting my time." To make matters worse, he lost his girlfriend—to a guy she'd met on Match.

Simon Glinsky, Kremen's friend and marketing consultant, saw Kremen's impatience with other people as part of what he called the "genius syndrome," his frustration that others couldn't solve problems as efficiently as he could. "Why hasn't this been done?" Kremen would snap, "this is so easy! It's just three steps. I could have done that last night." The engineers got so tired of his outbursts that they half jokingly put a strip of masking tape on the ground outside their offices and told Kremen not to pass it.

Finally the board told him they'd had enough. They told him they were going to replace him as CEO. Despite his insistence that Match was about content, they wanted to sell it as a software platform to newspapers. Kremen's stomach sank as he stared into the indifferent faces. They weren't just taking his baby away from him, they wanted to raise it to be a completely different person than he intended. And if that was the case, he told them, then he was out. And so, in March 1996, he resigned from Match.com without seeing a dime from his invention.

Kremen felt devastated. What had just happened? His head reeled. Life could be so cruel. Ever since he was a kid, he had tried to be his best. Sure, he had his troubles, his conflicts, he was far from perfect. But he had always been someone who strived, who bootstrapped himself from nothing, who dared to dream big, to be the Captain Kirk of his own *Enterprise*, to boldly go where no one had gone before. He had invented online dating not just to help himself, but to help the world, to find a new way to connect, to love, to be happy. And for what? To be ousted by a bunch of misguided suits? Kremen could feel the weight of his lost hopes crushing him. And in that moment he knew he had a choice: to let himself be buried in disappointment, or, better yet, begin something fresh.

And it was the latter path he chose. He even had a name for it: chasing the dragon—like crashing on one drug and getting wired on the next. He joined a startup as president; NetAngels used online software that suggested websites to visitors based on the person's traffic history. No one had seen anything like this before, essentially a recommendation engine for navigating the emerging world online. NetAngels, as a result, gained backers, including famed investor Esther Dyson and Ron Posner. Kremen, they recognized, stood out from the other innovators in Silicon Valley in a hugely important way: He was no one-trick pony. He was born with some weird knack for seeing around corners. Kremen made another $200,000 when he merged NetAngels with another firm, Firefly, and sold it to Microsoft. Kremen believed the best strategy for business was to place many bets, start ten companies in other words, hoping that maybe one will hit.

Kremen's brain was like a pachinko machine, a maze of silver thoughts bouncing between flashing lights. But when one would drop in the hole at the bottom, it would grab his attention and make him stop everything. And that's what happened when one particular silver ball landed in his mind: Sex.com. Kremen wasn't a pornographer, and had no interest in being one, but he was enough of an entrepreneur to know there was money to be made with that domain.

And then he thought of Stephen Cohen, the man who claimed to now own the site. Since they had last spoken, Kremen had been too busy dealing with Match to focus on getting Sex.com back from Cohen. But when he investigated it again, he discovered that Cohen hadn't been waiting around for him to catch up. In May 1996, Cohen had applied to trademark Sex.com, claiming, as he told Kremen, that he had been using the Sex.com name since 1979 on the French Connection. Furious, Kremen retained a lawyer to demand NSI provide answers to how and why did they hand over his site to Cohen in the first place?

Soon after, in mid-June, NSI forwarded Kremen the letter Cohen had sent them for the transference of Sex.com. Kremen couldn't believe his eyes as he held it in his trembling hands. It was typed on phony letterhead for Online Classifieds, Inc., the company name Kremen had used prior to changing it to Electric Classifieds when he registered Sex.com the year before, and it included his old home address as well as an invented, and grammatically incorrect, slogan: "For Your Online Ad's."

It was dated October 15, 1995, and addressed to Cohen, "re: Sex.com." Kremen read on with amazement:

> Dear Mr. Cohen, per our numerous conversations, we understand that you have been using Sex.com on your French Connections BBS since 1979 and now you want to use Sex .com as a domain name on the internet. Our corporation is the owner of Sex.com as it relates to the internet.
>
> At one time, we employed Gary Kremen who was hired for the express purpose of setting up our system. We allowed Mr. Kremen to be our administrative and technical

contact with the internet, because of his past experience with computers and their connections to the internet.

Subsequently, we were forced to dismiss Mr. Kremen. At no time, was Mr. Kremen ever a stockholder, officer, nor a director of our corporation and as the internet shows that Sex.com is listed in our corporation and not in Mr. Kremen's personal name. In fact, Mr. Kremen is the president of a different and unrelated corporation called Electric Classifieds, which is located at 340 Brandon Street in San Francisco, California. Further, Mr. Kremen's corporation owns Match .com which is listed with the internet registration.

We never got around to changing our administrative contact with the internet registration and now our Board of directors has decided to abandon the domain name Sex.com.

Because we do not have a direct connection to the internet, we request that you notify the internet registration on our behalf, to delete our domain name Sex.com. Further, we have no objections to your use of the domain name Sex.com and this letter shall serve as our authorization to the internet registration to transfer Sex.com to your corporation.

It was signed "Sharon Dimmick, President."

"Sharon Dimmick?" Kremen felt dizzy, infuriated, the blood rushing to his face as the letter spun through his head. Sharyn—not Sharon, as Cohen spelled the first name—Dimmick, a psychologist, was his roommate back in the early 1990s when he was living in San Francisco. She'd been living in the place before them, and they'd met through a mutual friend. The cohabitation had not gone well, with Dimmick frequently chastising Kremen for his dirty dishes, loud music, and allegedly letting her cat run loose. It

had gotten so strained, in fact, that Kremen had finally moved in with a lesbian couple downstairs just to have peace.

Seeing her name on the letter set his mind reeling. *"Sharyn fucking Dimmick?"* She wasn't the president of Online Classifieds, she was just a roommate he was living with at the time he'd registered the site. He had put their address down when he got Sex.com, so that would explain how Cohen would have known it. But why would her name be on this? Was that really her signature? Was she still pissed at him—she'd tried to break his arm after all—and now in cahoots with Cohen, and lying on his behalf? All the stuff in the letter about Kremen being hired to set up the computers, then being fired, it was bullshit, of course, since Online Classifieds was just him. Then he realized—her name was misspelled on the letter, Sharon instead of Sharyn.

And the most absurd thing of all? The assertion in the letter that a company called Online Classifieds would purport that they "do not have a direct connection to the internet," and so then would ask Cohen to delete their registration of Sex.com on their behalf. It was as if Julia Child said she didn't have her own kitchen so needed Cohen to make her a sandwich. If Kremen wasn't so furious, he would laugh. But this was no joke. Though he didn't have all the answers yet, there was one thing he thought for sure. Cohen had stolen Sex.com, and it was time to get it back.

CHAPTER 6

THE EROTIC TECHNOLOGICAL IMPULSE

"Almighty God, Lord of all life, we praise You for the advancements in computerized communications that we enjoy in our time. Sadly, however, there are those who are littering this information superhighway with obscene, indecent, and destructive pornography."

It was June 14, 1995, inside the Senate chamber in Washington, D.C., and Jim Exon, a seventy-four-year-old Democrat from Nebraska with silver hair and glasses, had begun his address to his colleagues with a prayer written for this occasion by the Senate chaplin. He was there to urge his fellow senators to pass his and Indiana senator Dan Coats's amendment to the Communications Decency Act, or CDA, which would extend the existing indecency and anti-obscenity laws to the "interactive computer services" of the burgeoning internet age. "Now, guide the senators," Exon continued his prayer, "when they consider ways of controlling the pollution of computer communications and how to preserve one of our greatest resources: The minds of our children and the future and moral strength of our Nation. Amen."

As the stone-faced senators watched, Exon held up a blue

binder that, he warned, was filled with the sort of "perverted pornography" that was "just a few clicks away" online. "I cannot and would not show these pictures to the Senate, I would not want our cameras to pick them up," he said, but "I hope that all of my colleagues, if they are interested, will come by my desk and take a look at this disgusting material."

They were interested.

One by one, they flipped through the pages of "grotesque stuff," as Coats put it, that innovation fostered. He cited figures—albeit dubious—from a study that found more than 450,000 pornographic images online that had been accessed approximately 6.4 million times the previous year. The main source had been the free newsgroups—alt.sex, alt.bestiality—and so on, that remained a Wild West of flesh and filth. "With old Internet technology, retrieving and viewing any graphic image on a PC at home could be laborious," Coats explained, forebodingly. "New Internet technology, like browsers for the Web, makes all this easier."

As urgent as the situation seemed to the senators, however, such concerns over pornography and emerging technology were far from new. John Tierney, a fellow at Columbia University who studied the cultural impact of technology, traced what he called the "erotic technological impulse" back at least 27,000 years—among the first clay-fired figures uncovered from that time were women with large breasts and behinds. "Sometimes the erotic has been a force driving technological innovation," Tierney wrote in *The New York Times* in 1994, "virtually always, from Stone Age sculpture to computer bulletin boards, it has been one of the first uses for a new medium."

Such depictions emerged, predictably, with every new technological advent. With cave art, there came sketches of reclining female nudes on walls of the La Magdelaine caves from 15,000

BC. When Sumerians discovered how to write cuneiform on clay tablets, they filled them with sonnets to vulvas. Among the early books printed on a Gutenberg press was a sixteenth-century collection of sex positions based on the sonnets of the man considered the first pornographer, Aretino—a book banned by the pope.

Each new medium followed a similar pattern of innovation, porn, and outrage. One of the first films shown commercially was the *The Kiss* in 1900, distributed by Thomas Edison and depicting eighteen seconds of a couple nuzzling. "The spectacle of the prolonged pasturing on each other's lips was beastly enough in life size on the stage but magnified to gargantuan proportions and repeated three times over it is absolutely disgusting," one critic wrote, while Edison celebrated how the film "brings down the house every time." The first erotic film, a striptease *Le Coucher de la Mariée,* released in 1896, was also heating up audiences.

By the late 1950s, the advent of 8mm film put the power of porn in anyone's hands—and launched the modern porn industry. When videocassette recorders entered the homes twenty years later, more than 75 percent of the tapes sold were porn. It became widely accepted that Sony's decision to ban porn from its competing Betamax format doomed it to oblivion. More recently, the breaking apart of the Bell phone system in 1984 spawned the explosion in 900 phone sex numbers. And so it was no surprise that the dawn of the internet was giving rise to the same kind of innovation, demand, and outrage that had been going on for eons.

The furor over internet pornography had started with the publication of a study, "Marketing Pornography on the Information Superhighway," in *The Georgetown Law Journal.* The authoritative-sounding study, written by a Carnegie Mellon undergraduate, Marty Rimm, claimed to be "a Survey of 917,410 Images, Description, Short Stories and Animations Downloaded 8.5 Million

Times by Consumers in Over 2000 Cities in Forty Countries, Provinces and Territories." Rimm asserted that 80 percent of images on newsgroups, the primary repository of pictures online, were porn.

The shocking figure caught the attention of *Time* magazine, which published a cover story on July 3, 1995, just in time for holiday readers, announcing the soon-to-be released findings. The cover photo showed a young boy at a computer keyboard, bathed in blue light, eyes wide, mouth opened in horror. "CYBERPORN," the cover line screamed, "a new study shows how pervasive and wild it really is. Can we protect our kids—and free speech?" As the writer put it in the piece, "If you think things are crazy now, though, wait until the politicians get hold of a report coming out this week."

He was right. Despite the outcry of civil libertarians and skeptics ("Rimm's implication that he might be able to determine 'the percentage of all images available on the Usenet that are pornographic on any given day' was sheer fantasy," as Mike Godwin wrote in *HotWired*), Rimm's study became the basis of the Communications Decency Act proposal. And, as Exon put it during the Senate gathering, their responsibility was clear. Despite objections over the restrictions on free speech, the CDA would target the burgeoning purveyors of porn online, who would now face up to two years in prison for posting obscene material that could be accessed by anyone under age eighteen. When the vote was taken, the answer was overwhelming: the Senate, and later the House, approved the CDA.

By summer, however, the basis of the law had been resoundingly discredited. Rimm's paper, savaged by critics, was found to have been published without peer review—feeding conspiracy theories that it was all the machinations of anti-porn activists. *The*

New York Times dismissed the study as "a rip-snorter," filled with "misleading analysis, ambiguous definitions and unsupported conclusions." Attacked by internet trolls, Rimm went into hiding. But his work, and the senators', was done.

On February 8, 1996, President Bill Clinton signed the Communications Decency Act into law. "Today," he said, "with the stroke of a pen, our laws will catch up with the future." For Exon and the others, it couldn't have come soon enough. "If nothing is done now," as he had urged his colleagues during the hearing, "the pornographers may become the primary beneficiary of the information revolution."

One day in Boca Raton, Florida, in May 1996, Jordan Levinson, the owner of AIS Marketing, a start-up that brokered ads for adult websites, received a call from a man who wanted to benefit from the burgeoning underworld of the information revolution: Stephen Cohen.

Levinson, who had worked with his father running a phone sex company, had encountered plenty of wannabe pornographers in his day, and sensed that, as he later put it, Cohen "didn't seem too knowledgeable of the industry." But Cohen had something more valuable: the one domain that anybody with a computer and a modem would type if they were looking for porn, www.sex.com. So he readily struck a deal to buy, sell, and collect ads for what Cohen promised, just as he had promised the board members at Sporting House, would be the single greatest destination for "sucking and fucking" online.

Despite the federal regulation, there was simply no way to stop the flood of porn online, let alone determine or enforce the age of consumers. And now more people than ever were online. According to the U.S. Census Bureau, the number of homes with computers was skyrocketing—approaching 36 percent of U.S. households, up from 22.8 percent in 1993, and just 8 percent in 1984. One in five Americans were now using the internet.

Of these, most reported using it for email or, as the Census Bureau catalogued it, "finding government, business, health, or education information," though anyone online at the time knew exactly what they were really seeking—just as generations had on every new medium before them. Even better, as Cohen learned, they were willing to pay for porn. When he launched Sex.com as a business in the spring of 1996, an underworld of outlaws, innovators, and entrepreneurs was racing to cash in. But first they had to do what no one had reliably done before: figure out how to make money online.

While Cohen might sell membership subscriptions to his site—charging visitors a monthly fee to access photos, videos, and so on—as Levinson explained, the trick was getting surfers to click a banner ad, the interactive billboards of the information superhighway, and visit a site. A banner ad on one page could be clicked and take a visitor to the other. The value for the advertisers came two ways: in "impressions," meaning the number of times the banners loaded up for visitors to see, and "clicks," the number of times someone clicked on the ad, which would take him to their site. "They pay for the advertising," as Levinson put it, "they pay for their banner spot to be there." Levinson would be his ad guy—buying, selling, and collecting money, all for a 15 percent cut. How much could Cohen get? With a site like Sex.com, Levinson thought, upward of $30,000 per ad.

Cohen didn't even need to make porn, he realized, to make money. He could just make money by selling ads on his site and cashing in on the traffic he sent to others. Cohen took one look at his blank webpage, and knew exactly what he wanted to do: sell as many banner ads as possible, and rake it in. All he needed to do was get the word out to the nascent pornographers online that he was open for business. And the place to do this was Vegas.

The burgeoning moguls and fans of internet porn gathered there for their annual convention, AdultDex, which coincided with Comdex, the annual computer trade show that drew 200,000 technology enthusiasts to town. Porn had long been a welcome draw at electronics shows since fueling the VCR boom in the 1980s. But times were changing. Two years earlier, AdultDex exhibitors got banned from Comdex for showing too much nudity—both on CD-ROMs and the scantily clad porn stars—in their booth (when the porn companies wouldn't leave, the Comdex organizers had to literally unplug their electricity to get them out the door). "Their stuff is obscene and we don't need them," a Comdex spokeswoman told the *Las Vegas Sun* after the convention in 1995, and if that meant losing the $500,000 in booth rental revenue, so be it.

But to the relief of the Comdex attendees, AdultDex refused to go for good. In November 1996, instead, they simply moved their computers and dominatrixes across the street to the Sahara, the Moroccan-themed hotel casino made famous in the 1950s by the Rat Pack. Cohen would be among the throngs streaming in and out of the porte-cochère entrance under the flashing lights of the yellow-domed minaret.

Across the smoke-filled casino floor, these were his people: the quick-talking moguls with fat wallets and chunky cell phones, the slinky adult actresses and actors at the slots, the wide-eyed Comdex attendees from Iowa with their laminate tags deftly flipped to hide their names. On the small exhibit floor, they showed off their software titles on smudgy screens: *The Dollhouse*, *Men in Motion*, *Virgins 2*. In another booth, a company was demonstrating

Showgirls Live, a live video feed, albeit painfully slow, that showed a stripper disrobing on-screen—an experience that could be had for $5 a minute. Jenna Jameson, a doe-eyed buxom blonde and the industry's most popular adult star, preened for photos as she extolled the wonders of her electronic mail. "It's so much easier than fan mail," she told a reporter from CNN, "let me tell you."

For Cohen, it was like being back at The Club in Orange County—the sex, the desire, the want, the money to be made, and the chance for him to be king. And he would be king of it all, he determined, because he had the most desired clubhouse online of all, the one that would be the first stop for the Iowans and their ilk online. He had Sex.com, and they would bow to him. Among those who were seeking Cohen out was Yishai Hibari, an Israeli musician turned adult webmaster who wanted to take out ads on the site. Word was that Cohen was doing three times the traffic. Among the porn stars in bikinis and guys with greased-back hair, he saw what he recalled to be "a funny chubby guy with an important look on his face," walking a small white Chihuahua with a red ribbon around its neck. Cohen was always chatty, and chummy, friends never saw him in a bad mood. "I heard you own Sex.com," Hibari told him.

"I don't know," Cohen replied, vaguely.

Hibari couldn't understand why he was being so circumspect. "Everything was unclear," he later recalled. But that, he learned, was Cohen's way, a tactic, however strange, for keeping people on edge and maintaining leverage. A few weeks later, Cohen relented—instructing Hibari to contact Levinson about buying precious space on his site—to the tune of $50,000 a banner. Sex.com wasn't pretty, Hibari could see, but Cohen's barebones use of it as a "banner farm" was a business coup. "It was genius," he said. Kevin Blatt, a marketing executive for adult sites, thought Cohen was, in his own way, visionary: someone who saw the value of the traffic, and realized the best way to cash in was by cramming as many possible banners on his site as he could.

It didn't take long for the success to go to Cohen's head. He became known for wandering porn shows with a smug smile, and a polo shirt embroidered with the Sex.com logo. Even among the rulers of the Wild Porn West online, he soon gained an unseemly reputation. Just as he had done for years, he put his amateur trademark attorney scam to use by suing anyone, and everyone, who had the word *sex* in a domain name. Serge Birbair, the owner of sexia.com, was among those who, as he put it, was "harassed by Stephen Cohen." When hit with Cohen's lawsuit, he didn't have the money to fight back against the traffic king—and chose instead to relent, and hand over sexia.com to Cohen. "It cost me money to defend myself, and it cost me a lot of grief," as one pornmaster put it after caving in. "Eventually, I decided it ain't worth the fight."

Cohen reveled in the power. No one could stop him with Sex .com on his side—not even the guy who claimed to own it. One day, Cohen received a certified letter from Kremen's attorney, demanding he not only cease and desist using Sex.com, but send the money he'd earned from the site Kremen's way. Cohen had one response. Kremen could go fuck himself. He'd been selling sucking and fucking online since the 1980s, and Sex.com was rightfully his. As he later told Kremen's attorney, "If anybody stole it, it was Gary Kremen stealing it from me."

"Welcome to SEX.COM! THIS IS A REALITY CHECK! Are you sure that you want to PROCEED to view explicit, sexually oriented materials? Further, Are you sure that you are over 18 years old, or of the legal age in your jurisdiction to view explicit, sexually oriented materials? (IF you do not know or are unsure, PLEASE now EXIT and contact an attorney in your jurisdiction for advice.)"

Kremen, hunched and bitter, fumed at his computer reading these words on the homepage of Sex.com. *This is my home, my domain!* he thought. *Cohen has no right to it!* Looking at it now felt like seeing someone had moved into his house and was sleeping with his girlfriend, if only he had one. It felt like the same old story, someone one-upping him, shaming him in the eyes of the world. He grabbed his mouse, and scrolled down the page like he was dragging nails down a chalkboard, scrolled by another warning. "SEX and SEXUAL ACTIVITY," it read, indiscriminately capitalized and misspelled like a ransom note, "including but not limited to NUDE MEN and WOMEN, SEXUAL PICTURES and SEXUAL ACTS, SWINGING and its lifestyle, NUDITY and its lifestyle, Gays, Lesibans [*sic*], Bi-Sexual Ladies, and light B&D. . . . EXIT by clicking the EXIT button below."

Past the EXIT button he scrolled, down, down until he came to the twenty-six-point all-capped bold font words like he was punching himself in the gut: "ENTER SEX.COM!" And he clicked. Kremen watched the next page slowly load down his screen: sluggish from the billboards of banner ads, one after the other, crammed over every available pixel. It felt like walking down 42nd Street and being assaulted by the porn marquees in 1977.

"Hardcore Sex" read one. "Come on in to our XXX Gallery" read another. One was a strip showing two naked guys sitting on the edge of a pool next to two nude women. There were enticements bordering a long column of links that read *ENTER*. "How to enjoy anal sex with your lover! the proper way!" *ENTER*. "A Ladies view on how to EAT PUSSY! the proper way!" *ENTER*. "DO IT! JUST CLICK HERE!" And then, at the bottom of the page, a contact email for any questions, Kremen read: Steve@sex.com.

Boom, there it was: the name, Steve@sex.com, Stephen Michael Cohen. Kremen rolled the name around in his mind, letting it embed, the tendrils of the enemy slithering into his brain's sulcus. Stephen Michael Cohen. The thief. The destroyer. He had no idea whether Cohen was actually making money from the site, but it didn't matter. There was principle. The man had stolen what was his. He had gotten screwed out of Match, he wasn't about to get fucked out of Sex.com.

Problem was, he was broke. Despite the payoffs of some investments, he needed more cash if he was going to file suit. While Kremen busied himself with new investments and consulting work, he hired a young attorney, Sheri Falco, to navigate the uncharted waters of a potential lawsuit. Falco, an intellectual property attorney, found Kremen doing a million things, as usual, in his office—starting an incubator, making calls, surfing the net. It didn't take long poking around for her to find out that Cohen was on a lawsuit tear of his own, riding on the back of the trademark protection he filed. She hit back, filing a trademark opposition to put his on hold.

Going on the public record against Cohen had another unintended effect. It got the attention of the many enemies he was making in the online porn industry. And, before long, Falco got a call from the two biggest, and most powerful, ones of all, Ron "Fantasy Man" Levi and Seth Warshavsky, who had an urgent message for Kremen. They wanted to help him take Cohen out.

They were good friends to have. Fantasy Man was considered by many to be the godfather of the online porn business. A dark-haired, imposing strongman, he lived in a ten-thousand-square-foot California mansion. Fantasy Man had been hustling since learning to shoot pool from the basis of the movie *The Hustler* himself, Fast Eddie, when he was just fourteen. With a knack for business and a passion for technology, Fantasy Man made his first fortune in audiotext, or phone sex, and parlayed that into the first large network of adult sites on the internet, Cybererotica.

His innovation: pioneering the traffic system of trading money for clicks that had since become the backbone of the business. It was called affiliate marketing, a kind of advertising system that would one day be used across the internet, from Amazon to eBay. Before long, he had more than one thousand porn sites in his lair, taking cuts every time someone clicked on an advertiser's banner, and becoming a multimillionaire. And one who was more than willing to flex his powers if anyone tried to cheat him out of his cash. "I made the rules on the net, because in those days it was like the wild, wild west," as Fantasy Man later put it.

While Fantasy Man was like the godfather, Warshavsky was the scandalous and diminutive boy wonder. A hyperactive twenty-three-year-old with a large nose and a snorting tic, Warshavsky had made his first fortune in his teens, running a phone sex business out of his bedroom in Seattle. A computer geek, he immediately moved his empire on the early net, establishing a company, the Internet Entertainment Group, that pioneered the most electric innovation in the online porn world, online peepshows.

For a price, visitors to his main site, Club Love, could chat with naked men and women for about $1 a minute. "This is it. This is huge," he told a reporter for *The Wall Street Journal*, which ran a front-page story on his booming business. "This combines the interactivity of phone sex with the visuals of television." The *Journal* marveled at how "Cyberporn is fast becoming the envy of the Internet. While many other Web outposts are flailing, adult sites are taking in millions of dollars a month. Find a Web site that is in the black and, chances are, its business and content are distinctly blue."

The article detailed how the innovators in porn had done more than just slap dirty pictures online. Pornographers like Fantasy Man and Warshavsky had figured out nifty innovations in internet marketing, created ads that "popped up" in front of webpages, and succeeded in upselling visitors to actually shelling out money for subscriptions. They'd also created new kinds of delivery mechanisms, secure credit card payments, and live videos. "Internet pornographers deploy savvy tactics that mainstream sites would do well to imitate," as the story put it. Bob Guccione, the publisher of *Penthouse* who represented the old media sex empire, put it succinctly to the *Journal* when he said "There are a lot of computer nerds emerging as porno kings."

And, for that matter, porno queens. Women were among the most innovative and successful entrepreneurs in the business. Beth Mansfield, an Army brat and NASCAR fan from Alabama, was a single mom and unemployed accountant living in a mobile home when she heard of people making money in porn online. Mansfield didn't want to make porn, however, so she began carefully curating pages of links to other sites. A concerned mom, she refused to use profanity on her pages, and would substitute asterisks to cloak sh*t and f*ck. But perhaps her greatest innovation was branding—naming the site Persian Kitty, after her cat. Something about the mystique of the name, the idea that a woman was behind the site, went viral—even more so because Mansfield kept her real identity anonymous. Before long, she was selling ads across the web to sites that paid a premium to be listed on her page. In her first year, she made $3.5 million.

A few miles from Mansfield's mansion in Seattle, an ambitious young stripper named Danni Ashe read a book on HTML programming during a beach vacation. She launched her own fan site online, *Danni's Hard Drive*, in 1995 as a place to put her own promotional pictures. Then Ashe struck on a more lucrative idea—charging for membership, still a new idea at the time. She hired models, posted pictures, audio interviews, videos, and charged $15 a month for access—becoming one of the first subscription sites on the internet, besides *The Wall Street Journal* (which later profiled her in a page-one story about online pornographers, called "Lessons for the Mainstream"). Before long, Ashe was making $2.5 million a year and reportedly using more bandwidth than all of Central America.

As the moguls of porn became the envy of the internet, the federal government conveniently got out of their way. On June 26, 1997, after more than a year of heated debate about the censoring of the internet, the United States Supreme Court struck down the Communications Decency Act for violating the First Amendment. It was a landmark decision, protecting the young medium from government regulation. For better or worse, online porn was here to stay. As humorist Dave Barry put it a few months later, after visiting that year's AdultDex convention, "this fast-growing billion-dollar industry will undoubtedly come up with newer and better ways to help losers whack off."

But Fantasy Man and Warshavsky, as Kremen was told, had one big problem: Cohen. Fantasy Man had been introduced to Cohen already at a trade show, and found him boorish. But more recently he'd been hearing bad things from his friends in the business. One by one, he learned, they were getting sued by Cohen. He didn't like that at all. *Who the fuck did this guy think he was?* When his friend Serge Birbair called him after he'd been sued, and told him that Cohen, supposedly, had stolen the site from some penniless nerd named Gary Kremen, that was all Fantasy Man needed to hear.

He picked up his phone, and called Warshavsky—proposing that he be the Robin to his Batman in a fight against Cohen. "I want to get ahold of this guy Gary," he told Warshavsky, "who's saying that he owns Sex.com and Cohen stole it from him and help him with the lawsuit, because he doesn't have any money for a lawsuit. Wanna join in with me?"

Warshavsky, at the time, was riding higher than ever. He had recently released a stolen sex video of actress Pamela Anderson and her husband, Tommy Lee, on his site, driving legions of web surfers to his site. When he heard Fantasy Man's proposal, Warshavsky snorted approvingly. He didn't like the way everyone was being bullied by this guy either. "They were intimidated," as Warshavsky later put it to *Wired*. "He basically strong-armed them. The guy is really kind of a scumbag."

But this wasn't just altruism, it was about money. After some due diligence, they determined that Kremen, the Match.com inventor and Stanford MBA, had a credible complaint. So they wanted to get on board. When they called Kremen, they could hear the anxiety in his voice. "He seemed like a tech nerd," Fantasy Man recalled, "antisocial. A guy not interested in adult." But a smart guy nonetheless. Kremen felt a mutual respect, intellectually at least. Warshavsky seemed like a weird live wire, but Fantasy Man was pure business.

They told Kremen what they had in mind. They would put up $100,000 money for him to sue Cohen—in exchange for 51 percent of Sex.com if and when he got the domain back. Kremen wasn't interested in running Sex.com, he didn't aspire to be some kind of twenty-first-century Larry Flynt. He just wanted justice. So if these two guys wanted to help him in exchange for a controlling interest in Sex.com, so be it. "That's great!" Kremen told them.

With a few grand from Fantasy Man and Warshavsky, Kremen got Falco to dig in and turn up whatever she could against Cohen. They sent a mass email across the adult industry online, asking for information from anyone who'd been targeted by his enemy's wrath. "Several sites, such as Sexia.Com, SexCom.Net, HotSex .Com have been either threatened or actually sued by Mr. Cohen for trademark infringement for the use of the word Sex in their domain name," he wrote. "We are interested in obtaining information on this matter and are willing to share what we know about Mr. Cohen's actions."

But he also had to reach out to the alleged author of the letter that was used to get Sex.com from NSI: his estranged ex-roommate, Sharyn Dimmick. Falco found her living in a trailer north of San Francisco, and Dimmick was unequivocal when shown the letter. It was forged, she said, and signed an affidavit verifying as much, as well as that she didn't have the authority to transfer the rights in the first place. Affidavit in hand, Falco fired it off to Network Solutions, along with another request to give the site back.

While Kremen waited for a response from NSI, he used the leverage to go after Cohen directly too. But Cohen, he discovered, was busy covering his trail. With the help of his attorneys, Kremen learned that Cohen had sold Sex.com to a Mexican company called Sand Man Internacional, based in Tijuana. And Sand Man was owned by another company, Ocean Fund International, based in Tortola, British Virgin Islands.

In a press release by Ocean Fund, the chairman, the regally named Sir William Douglas, said Sex.com was bringing in a net income of $95.5 million in the fourth quarter alone. In fact, the site was generating so much traffic that it was causing "major internet congestion problems," the release stated, which forced Sex.com to open its own internet network access point in Tijuana. With traffic growing at 18 percent per month, Sir William Douglas went on, Cohen was worth every penny of his $17 million salary and

$100 million in stock options. “I have full confidence in a smooth transition of our roles and that Stephen will succeed in bringing further growth and accomplishments to our Company,” Douglas stated.

This was just a shell game, Kremen suspected, phony companies made by Cohen to keep his money safe. Cohen’s lawyer made their position perfectly clear during a call in February 1998 in which he told Falco that Kremen was “a kook” who “would never see a penny” of what he was asking because Cohen’s companies were incorporated offshore. A few weeks later, NSI followed suit—denying Kremen’s request to get back the site and, effectively, leaving it up to the courts to decide.

Kremen felt the air leave his lungs. He had done everything right, he thought: been in the right place at the right time, registered domains when no one cared, started an online dating empire, and for what? He felt defeated. And it got worse when, over his objections, the Match.com board decided to sell it to Cendant, a consumer-services company in Connecticut, for a paltry $7 million. Despite pioneering online dating, Kremen left with nothing but a title on his Match.com profile, “Founder.”

But he wasn’t going to lose out again. Cohen hadn’t just stolen his website, he believed, he’d taken his property. And the implications of that went far beyond him. He wasn’t just fighting for Sex.com, he was fighting for the future of the internet, for the soul of what it meant to stake a claim in this new world online. The internet wasn’t just some invisible world of ones and zeroes, it was a place, a world where humanity would soon migrate. If someone could just take another person’s property, then what would that mean for the future of life online? It would mean that it meant nothing.

And so, on July 9, 1998, Kremen sued Cohen in the U.S.

District Court for the Northern District of California, "seeking damages, injunctive relief and a declaration of ownership for an Internet domain name." Soon after, he added NSI as a defendant, alleging that they were liable for transferring the domain name based on Cohen's forged letter. Then he steadied himself. As much as he thought it was everyone's war, it was his alone to fight.

CHAPTER 7

THE ART OF WAR

There was the Switch Hitter, the Double Whammy, the Anvil Stroke. And then: the Shuttle Cock, the Bookends, the Flame. Moreover, the Base Clutch, the Love Tug, the Two-Timer, the Thigh-Swatter, Best Fist Forward, the Milker, the Perpetual Penetration, the Palm Swirl, Tiny Circles, the Ring, the Door Knob, the Shaft, and the Small Pinch. They were all the names of the techniques listed by Stephen Cohen in the instructional essay he was writing for Sex.com on "How to Give the Perfect Handjob." He titled the essay in hot pink font against a bright yellow background, and continued in italicized black. "Sex means more than intercourse, exploring all the different variations enhances your sex life and keeps it from getting stale," he typed. ". . . So, read on and learn how to let your fingers do the walking."

Cohen had reason to feel excited—and not just because he'd won $86,000 in Vegas on a Caribbean poker machine. By the summer of 1998, running Sex.com had become the central passion of Cohen's life. It had delivered on the promise he'd only imagined back when he launched the French Connection BBS, creating an online empire of sex for which he alone would be king. It also felt like validation for just how insanely far ahead of the curve he really was, despite his run-ins with the law: it was him—not Bill Gates (whom he claimed to have hobnobbed with at a computer convention), not Steve Jobs, certainly not Gary fucking Kremen—who realized that the real engine of the internet was the two things everybody wanted even if they wouldn't admit it: the sucking and the fucking.

And he had once again found love. Her name was Rosa Montano, a secretary at a bank he'd met through a mutual friend in Mexico, where he'd been traveling for years. He affectionately called her "Rosey." Rosey, who had two daughters of her own, came from a large family, with five brothers and sisters, and was just the kind of woman Cohen needed: one who provided the sense of family he'd long desired, but left him to his business without asking questions. In November 1997, he married her in Vegas—his fifth marriage but, she hoped, his last.

Since then, he'd been splitting his time between Vegas, where his mother still lived, and Mexico, where he moved with Rosey and her kids into San Antonio Del Mar, an oceanfront community thirty minutes southwest of Tijuana. He opened an office on a busy street in Tijuana, where he ran the day-to-day action of Sex.com. Cohen had help in Marshall Zolp, whom he'd met in prison. Zolp had a history of security fraud, including a $100 million pension scam, as well as bilking investors out of $2.4 million in a phony company that claimed to have invented a self-chilling beer can. Zolp oversaw the live video feeds that ran over Sex.com, arranging for the women in Tijuana to undress in front of a flank of computers. "No per minute charges!!!" Cohen wrote on the website. "No Software Needed!!! Pick the girl(s) you want to watch from any of our 11 hot shows! Some stages also have FREE CHAT, where you call the shots! And our MAIN STAGE features fluid motion video with SOUND."

While Zolp handled the strippers, Cohen badgered Jordan Levinson, his marketing consultant, about the banner ads, which were going up and up in price with demand. With nine million members, he told buyers, he was claiming upward of 146 million unique hits daily. A single banner ad could sell as much as $50,000 a pop, with some advertisers spending more than $1 million per month alone. "The entire world was typing Sex.com," said Jonathan Silverstein, who worked with Club Love, calling it a "banner farm that was printing money."

Analysts valued the site at more than $100 million, but Levinson tried to appease his boss with even bigger numbers. He denied his better instincts when Cohen wanted to redesign the site so as to cram more banner ads on a page, despite the garish design. But Cohen wasn't out to win any interactive art awards. He only wanted to continue doing the one thing that everyone in the online gold rush was dreaming about: making millions through electronic commerce online.

With so much money and traffic coming in, Cohen hatched an even bigger plan for his empire, to make his company worth $60 to $100 billion. As he saw it, domain names were just the beginning. The bigger money in the future was going to be about bringing internet access to the masses by building out an infrastructure—cable, microwave, and so on—that's where the money would be, he explained:

> An example, AT&T just purchased TCI, which is a cable company in Colorado, paid $40 billion for it, and it's a complete fiber infrastructure. Sex.com, if it becomes part of this, will be able to provide a 24/7 production on this—on this

fiber system, which will substantially raise its value, and it will raise the value of the whole thing. But the real value is not going to be in the domain names in the future. The real value is going to be in the infrastructure. This month the Internet authority is just about to release four or five other domain names which are going—which are going to even diminish the value of Sex.com.

Another thing that you've got to take into consideration is that in a lot of foreign countries, sex is something that is not readily available. A lot of people are married through their parents, through arranged weddings. Sex is not something that is free and easy as it is in the United States and some foreign countries. And we get a lot of users from these areas, because they're basically sexually frustrated. As society [as] a whole starts to enlarge and freedoms become more available, the value of Sex.com starts to diminish. The fiber infrastructure is the only thing left that will keep a real true value in Sex.com. And it becomes a very true value.

I have a different concept than most people. I believe that the whole Internet is going to change. I believe that telephones that we know today are going to change. I believe that television as we know it today is going to change. And I think it's all going to be IP to fiber optics. I believe—and my company believes—that the people that own the fiber optics and the infrastructure are going to be sitting in the perfect position. I think Sex.com will be more of a transmission in the future over a TV over IP. I think the rights could be worth upper billions.

So Cohen set about building his new dream: becoming the largest internet service provider to Tijuana. Cohen had long been an engineering geek at heart, not unlike Kremen, someone who relished the feeling of a tool in his hand, a screw to be tightened. So while the computers buzzed in his warehouse on Diego Rivera Avenue, he got to work. Just beyond the Border Patrol in San Ysidro, California, and up a hill behind a train track on a short dirt road called Rail Court, he found a perfect patch of land on which to build a point of presence, or POP, from which he could build his internet service provider. The land had once been the location of a used car lot, but had since fallen into disrepair. A fence surrounded a patch of withered palm trees, and a dilapidated three-room, one-story shack in the middle.

Cohen headed up there one dusty afternoon past the railroad tracks, and purchased it for $500,000. Needing to get the rights to run the fiber optics down to the border, Cohen had to strike a deal with the railroad, which hired an assessor to survey the metes and bounds. With his plan approved, Cohen and several Mexicans who worked for Sand Man came out one day to set everything up. As he stood under the hot sun, he gazed out over the buildings of Tijuana in the distance, imagining the bright future to come. *Bueno.*

But while Cohen broke ground on his internet company, he was still having to contend with swatting Kremen away. The stakes couldn't have been higher. Sex.com was the basis of everything, and he couldn't afford to lose it. And more, NSI had notified him that they were conducting their own internal investigation into the domain name's registration; if Kremen's allegations in his lawsuit were true, NSI went on, not only would they return Sex.com to Kremen, but seek indemnity from Cohen as against any liability from Kremen's suit.

But the more Cohen read up on the case, the more confident he became. Cohen had noticed that Kremen had filed his complaint by naming his old company, Online Classifieds, as a co-plaintiff.

But when Cohen searched an online database, he came to a startling realization: Online Classifieds hadn't even been incorporated at the time Kremen registered the Sex.com domain. This meant, he concluded, that he could have the case dismissed on the grounds that Online Classifieds didn't even exist. Whoever owns this domain isn't Gary Kremen, Cohen's attorney, Bob Dorband, could argue. In fact, it wasn't until just before filing the lawsuit against Cohen that Kremen had his attorney incorporate Online Classifieds. Cohen even got the documents to prove it. And so, with his attorney's help, they filed a motion to have the case dropped.

Cohen flexed his confidence when a reporter, Craig Bicknell from *Wired*, called him to do a piece on the battle over Sex.com. "Let me make it real simple for you," Cohen told Bicknell, "our audience is not America. It's the whole world. There's only one word in the whole world that everyone understands—sex. You type the word 'sex,' you come to Sex.com."

When the *Wired* article came out, on April 15, 1999, Cohen got what he called "an emergency call" from Levinson to check out the story. Cohen's eyes skimmed over the words. He was relieved to see that the NSI spokesperson wasn't taking sides, given the company policy of letting disputants sort out their own resolutions. "It's up to the two parties to work it out," the spokesperson said. But the deeper he got into the story, the more alarmed he felt. Sex.com, *Wired* wrote, was "a blindingly garish site that would put pre-Disney Times Square to shame . . . if even half the allegations that Kremen makes are true, the tale behind Sex.com is the most sordid in the short history of the Internet economy."

Cohen's enemies were quoted anonymously speaking out against him. One webmaster dismissed Sand Man and Ocean Fund as part of his con. "The companies don't exist, except in name," the webmaster told *Wired*. "He created these shell corporations with

P.O. boxes in other countries, so it's a real pain in the ass to serve him with subpoenas." Kremen agreed, suggesting that Cohen's move was a way to keep him from getting Sex.com back. "It's to make it hard to go seize it," he said. "It was stolen, literally stolen," Kremen went on. "The case is simple, it's about an international con man, who was twice in jail, forging a letter . . . and taking away a domain name—and Network Solutions doing nothing about it."

By the time Cohen finished the piece, all he felt was one thing, as he would soon tell the court: rage.

"Will you state your full name, please?"

"Gary Alan Kremen."

"Mr. Kremen, how old are you?"

"I'm 35 years old."

"Have you ever been known by any other name?"

"Idiot. Fool. Hey You, et cetera."

It was April 19, 1999, inside the conference room of a fancy law office in San Jose, California, where Kremen was being deposed in his case against Cohen and Network Solutions. Nervous and uncombed, his paunch wedged against the shiny round conference table, he struggled to remain calm as he fielded questions from Cohen's attorney, Bob Dorband, with NSI's two stoic lawyers and a videographer with a giant camera zooming in close. His effort to lighten the mood with his wisecrack about his aliases fell flat. "Have you ever given testimony in a case before?" Dorband asked.

"No," Kremen replied. But in the interminable eight months since he'd filed the lawsuit, the case had been taking its toll. "It's given me an upset stomach and headaches," he explained, "and my back hurts, and my spleen hurts too." The stress was getting to him, he admitted. He was taking Trazodone to help him sleep,

Paxil for anxiety. This seemed to explain why Kremen looked bug-eyed and bleary, his hair slick from sweat. Dorband, sensing an opportunity to discredit Kremen, asked if he was "taking any drugs that might impair your memory?" Kremen was honest to a fault. "Maybe," he said.

Tracking Cohen was taking up more and more of his time and energy: finding out how much money he was making, how many ads he was selling, how he was profiting from what was rightfully his. He'd surf over to Sex.com, and feel his gut twist every time he saw another banner ad on the filthy homepage. It wasn't just the loss of business that was eating away at him, it was his burning need to prove himself again.

Since leaving Match, he'd been on overdrive: investing in start-ups, and helping other companies buy and sell domains. There had been successes and riches. With the boom of the net, NSI was now registering nearly 200,000 new domains a month, and more and more companies were paying a premium for names. Kremen helped AltaVista, a search engine, sell its domain to Compaq for $3.3 million, and brokered other deals for computers.com and coffee.com.

He'd garnered the attention of *The New York Times Magazine,* which featured him in a story about the domain name battles called "Greed Dot Com"—and questioned the honorability of the trade. "If you're not using a domain name yourself, what about just giving it back?" the story asked. "Giving it back?" Kremen replied incredulously. "It's property! That's like saying, I'm not going to use this piece of land, I'm going to give it back."

All the while, the bile of the Match.com fiasco had only grown worse. Cendant, the company that had purchased the site the year before, was now about to sell it to Ticketmaster for $50 million. But Kremen, who no longer had a stake in the business, didn't make any cash. His idea. His site. His invention. Gone. And why? Because he'd tried to do the right thing, keep them from ditching it too soon, opening it up to the gay and lesbian market, and more. Even worse, he felt like everyone in the Valley seemed to know. "When you start a company, everyone knows it," he told the *Mercury News* for a story about the struggles of dot-com moguls. "It made me a little bit gun-shy. It's a type of performance anxiety."

It wasn't just his professional struggles that were in the public eye, it was his personal ones too. The news stories about online dating were still billing him as the lonely heart who gave love to a generation but still hadn't found it himself. "It's really a bummer," Kremen told the *San Francisco Bay Guardian*, which included a photo of him looking forlorn, holding a dozen roses. "I've helped a lot of people but it just hasn't helped me." For all these reasons, he was even more determined to wrest Sex.com back to his control again. The site was "a key cornerstone property in one of the biggest areas of the Net," he said, easily worth $100 million, or more. And he didn't need to be a degenerate to cash in.

As he told the attorneys, he had already written up a business plan. He was not about to become a pornographer. Instead he was going to take what he called a "softer" approach to Sex.com. The site, he intended, would be educational, Kremen explained, similar to other successful health and wellness sites appearing online, such as *Dr. Koop* (the former surgeon general's site, which had received more than a billion dollars in capitalization). Kremen would take a "Joycelyn Elders approach instead of the Larry Flynt approach," as he put it. "What I could have done with it is made it a portal, as I talked about, for kind of health-related information, sex—positive information that would be accepted as very good content by several companies that are now publicly traded."

Kremen bristled when he was asked if, instead, he'd simply tried to register any other sites after he "lost" Sex.com. "I didn't lose it," he said. "It was stolen." All that mattered was that Cohen had robbed him, and he was pissed off. And the more they pressed him, the more he pushed back—likening Cohen's actions to those of "an international con man" and "criminal mastermind." He suggested he had not only stolen the site, but doctored old records of the French Connection to make it appear like Sex.com existed decades before the dot-com nomenclature even existed. "NSI should have known that," he snapped, "they controlled them all. They were grossly negligent."

"You consider Mr. Cohen to be a criminal mastermind?" Dorband asked.

"Criminal, yes," Kremen said, "mastermind, we'll see."

And if he was facing a criminal mastermind, Kremen resolved after the deposition, then he was going to need more help fighting back against him. A friend suggested she knew just the right attorney for him. So one day in the summer of 1999, Kremen drove his beat-up old Honda down to Ojai to meet Charles Carreon, a long-haired, leather-jacketed, motorcycle-riding, forty-three-year-old attorney who seemed just wild enough to take on the wildest case online.

Like Kremen, Carreon had spent his life looking for and living on new frontiers. Raised in a strict Mexican Catholic family in Arizona, Carreon escaped by becoming a full-blown child of the 1960s—dropping acid, joining the apocalyptic cult Children of God for a time, and backpacking Europe, before fleeing back to America to pursue Buddhism and law school. Since getting his law degree at UCLA, Carreon had been defending small-time drug dealers in rural Oregon, but had moved to Ojai with his family for a fresh start—one that the I Ching had not yet revealed for him.

It didn't take long after talking to Kremen over a lunch of Hawaiian pizza to feel like the universe had put a new path before him. As Kremen told him the story of his battle against Cohen, Carreon took an instant like to this uncombed oddball from Silicon Valley. He appreciated what he later described as Kremen's "mischievous, nerdy but huge energy," the way his mind fired in a million directions a minute. But he could also sense that Kremen was feeling utterly beaten by Cohen, and desperate for a way to win. "You didn't perceive the value of Sex.com initially," Carreon told him, philosophically. "It's a very bad habit to allow people to take things of great value away from you and not fight until you lose or win. It is a much better solution to lose that battle having fought it."

Kremen felt a kinship with Carreon's fighting spirit, the way he twirled his chrome-handled lock knife while he spoke, and freely quoted from the tattered copy of Sun Tzu's *The Art of War* he carried everywhere in his briefcase. "I have heard of some campaigns that were clumsy and swift," as Carreon quoted, "but I have not heard of any campaigns that were lengthy and skillful." Kremen thought he was some kind of hippie genius, and, more important, someone who gave as much of a shit about this case as he did. After all the madness, Kremen had become wary of trusting anyone—personally or professionally—but Carreon broke through. *Finally*, Kremen thought, *somebody cares*.

And when Kremen described all the money to be had, Carreon, who'd been struggling to get by, widened his eyes. "This is the case I've been waiting for," Carreon told him. In return for $100 per hour and 15 percent of Sex.com when Kremen got it back, Carreon agreed to work for him full-time. With a stake in the most valuable site online, he figured he'd soon make millions.

The two quickly became inseparable, a gonzo duo like Hunter S.

Thompson and Oscar "Zeta" Acosta in *Fear and Loathing in Las Vegas*. Carreon followed Kremen back on his motorcycle to Haight-Ashbury, where Kremen was now living in a second-floor walk-up with the smell of pot wafting up through the window. Long into the night, they stayed up poring over the case. Kremen fumed over how he was being portrayed by Cohen and NSI as if he stole Sex .com, when all along it was his.

Carreon twirled his pocketknife, and tried to put him at ease. Cohen's attorney, he told him, was just practicing what he called "smallpox blankets and firewater," like when lawyers conspired to steal Indian land with biological warfare and fake documents. "That's how the West was really won," Carreon said, "it was sneaky and dirty. Like the old rail barons, Cohen had produced a phony deed to steal Sex.com." This is why Kremen had needed to hire someone with "Oregon frontier savvy. Muleskinner wisdom," as Carreon put it. "Someone like me."

Up until then, Carreon determined, Kremen had been keeping too low a profile with his battle for Sex.com. When Kremen explained that he didn't want to be perceived as a pornographer, Carreon pushed back. By acting that way, he said, he and Cohen just looked like "two junkies fighting over a dime bag," as he put it. "Your average judge or juror might throw up his hands and just say, 'Who cares? It'll kill you both!' " Carreon advised a more philosophical tactic. "We have to escape the stigma that attached to sex itself," he said.

He fished around the apartment through the strewn *Wall Street Journal*s and *Barron's*es to find Kremen's handwritten, two-page Sex.com business plan. "We have to embrace what everyone has overlooked," he went on. "Gary Kremen, the Stanford MBA and Internet visionary, the originator of Match.Com, the world's largest matchmaking site, would have developed Sex.Com as," he quoted from the business plan, "a 'public health, woman-friendly site' à la 'Dr. Ruth' or 'Dr.Koop.Com.'" The words sounded like music to Kremen. It was a fantastic idea. *Match gave credence to the entire idea of a "wholesome" Sex.com,* he thought.

But that wasn't all, Carreon went on. If they were going to embrace the idea of a kinder, gentler Sex.com in order to help win it back, they had to dig in their heels on a bigger idea, one that Kremen had been evangelizing but no one was yet realizing: this was a fight for property rights. The price of online property, after all, had only been skyrocketing: WallStreet.com had recently sold for $1 million, Business.com for $7.5 million.

As Carreon pointed out, California law defined property as "everything capable of being owned." Sex.com not only fit the bill, it was what Carreon called "a property magnet." Every credit card subscription that Cohen was processing, Carreon explained, represented property that Kremen rightfully owned. "You do the math," Carreon told Kremen. "That's a lot of money for a country boy."

And, Carreon went on, pursuing the property claim would bolster their case against NSI too. "If Sex.Com was property," he later wrote about the case, "and Gary was the owner by virtue of being the first to register the domain name, then NSI should have some obligation to protect his property from being transferred to another person without his permission." Therefore, the stakes—not just for Kremen, but any future prospector online—were immeasurable. "Being the most valuable domain," he went on, "Sex.Com presented the best-case scenario for a judicial finding that domain names are in fact property. The eyes of the world would be upon this case."

Even though he was the spokesperson for the world's largest hotel owner, Starwood Hotel and Resorts Worldwide Incorporated, Jim Gallagher wasn't accustomed to getting woken up in the middle of the night with an urgent call from the press. But that's what happened at 1 a.m. on June 15, 1999, when a reporter from the *Las Vegas Sun* rang, wanting to know if the press release was true. Gallagher gripped his phone. Two months earlier, Starwood had agreed to sell its eight Caesars Palace properties to Park Place Entertainment for $3 billion, but there was nothing new to report. What press release? Gallagher wanted to know.

"Internet Porn Giant to Buy Caesars Palace," read the title of the item that had just hit the PR Newswire. "Ocean Fund International, a British Virgin Island Mutual Fund and the owner of the world's largest pornographic Internet site, SEX.COM, today submitted a $3.6-billion, all-cash offer to purchase the eight Caesars Locations recently purchased by Park Place Entertainment." The company pledged an additional $2.5 billion for capital improvements, including plans to build a sports arena at Caesars Palace in Las Vegas.

The move was just the latest in the company's expanding empire, the press release touted. "There are some people in this industry who claim to make a lot of money," Ocean Fund's chair, Sir William Douglas, stated. "In fact, SEX .COM does 86% of the sex-related business on the web. We make more than *Penthouse*, *Hustler*, *Playboy*, and all other major sex web sites combined. I challenge anyone who claims to do more business in this industry to produce auditable numbers." The company's wholly owned subsidiary in Mexico, Sand Man Internacional, was already investing $100 million to bring free fiber optics to every resident and business in Tijuana. "We are extremely proud of our demonstrated record of success," Sir Douglas concluded. "We will do no less with the addition of Caesars to the Ocean Fund leisure and luxury-industry portfolio."

But as Gallagher groggily explained to the reporter, he had no idea what this was about. Despite what the release claimed, he went on, Starwood hadn't received an offer—or even heard of Ocean Fund at all. They weren't the only ones mystified by the press release, which was riddled with typos and errors (Caesars' riverboat, "The Glory of Rome" was misspelled "The Glory of Roam"). Neither Ocean Fund nor Sand Man was listed with the Securities and Exchange Commission. "We'd characterize it as bizarre," said an analyst at Bear Stearns. "Our advice to investors is to dismiss it as noise."

When reached at his Salt Lake City office, Bob Meredith, the lawyer for Ocean Fund, remained just as circumspect about the company's offer. "I don't know how much thought they have put into it," said Meredith, who could not provide more detail on the company's board than what was published. "Douglas?" he told a journalist at the time. "He's some cat in the Virgin Islands is all I know."

By the next day, the offer had unraveled into what Gallagher called "a total sham." Sir William Douglas, the alleged chair of Ocean Fund, denied any association with the company, prompting an apology from the British newspaper *The Independent* for publishing statements from the press release that were "incorrect and unfounded." The bid was concluded to be little more than a publicity stunt—but an effective one at that. Within twenty-four hours, Sex.com had generated headlines from the *Los Angeles Times* to the *New York Post*, which surmised that "Wall Street hadn't seen such cojones in years."

It also managed to move the market, boosting Starwood's stock by 44 cents, and dropping Park Place's by 19 cents on the New York Stock Exchange. "After following gambling stocks for 10 years," the Bear Stearns analyst said, "you think you've seen it all. Then something like this comes along that makes you shake your head." But in addition to promoting Sex.com, it also added to the notoriety of the "porno peddler," as the *Post* put it, who was revealed to be responsible for it all: Stephen Cohen.

Cohen's antics weren't just garnering the attention of the mainstream world. He was now among the most infamous "peddlers" in the net's most notorious industry—and at war with just about everyone in the online sex trade who crossed his path. By the summer of 1999, Cohen's battle with Kremen, and Kremen's high-profile backers, Fantasy Man and Warshavsky, had made him the industry pariah.

But he had other concerns: building his ISP, and running Sex.com. He spent his days zipping back and forth across the border to his fiber optics hub. It became so routine that Cohen grew weary of the red tape he had to deal with every time he passed through Customs. Instead of stopping to claim whatever duty-free items he had, he'd just reach out the window and put a $100 bill in the border agent's hand and keep going. If he didn't like the agent, he'd simply reverse backward on the highway, and split.

Sitting back in his office in Tijuana, Cohen readily spun his story on the phone to anyone who'd listen. "When Kremen filed this lawsuit, he did us a big favor," Cohen told Luke Ford, who wrote a widely read blog about the online porn business, "because if you don't sue someone within a period of time, it can be construed as giving implied permission. We decided at that point, instead of fighting forty lawsuits by people using the name Sex.com, to take care of Kremen." This included both Fantasy Man and Warshavsky, whom Cohen considered part of Kremen's team. "We've got Warshavsky's tit in a wringer," he went on. "We've got him cold. He's fucked."

Cohen slipped into his pedantic mode, deriding the other dot-com moguls as technological neophytes overinflating their claims. "We have 8,799,232 members," he told Ford. "The other adult internet sites are not even in the ballgame. Most people do not understand the internet. You get these fools like Seth Warshavsky who claims he's doing all these millions of dollars' worth of business. If writers were more technical they could check out the information."

Cohen told Ford to do a WHOIS search for Sex.com, and waited as he heard Ford's fingers rattle across his keys. He wanted to show Ford proof that he was dominating Warshavsky's traffic. "Do you see the number 11083?" he asked. "That's an autonomous system number. That means you report to more than one place. That means that more than one ISP feeds you." The implication was that Sex.com needed more than one internet service provider to handle all its eyeballs.

He had Ford then type in the address of Warshavsky's site, Club Love. "Hear that noise in the background?" Cohen asked, as he put the phone up to his own computer, which played back a fuzzy sound like TV static. "Club Love's internet provider is cable and wireless," he went on. "He's running out of an IP address of 166.48.217.250. Club Love is registered as JNS Communications Inc. and the address is 208.139.0.21. He only has one ISP. He's probably running less than a T3 line, and yet he claims he's doing all this business. From what I can see he's not running more than a couple of T1s. His claims of millions of dollars are all BS.

"Now, let's take a look at Cybererotica. Cybererotica is fed by IGallery. Cybererotica is running greater than T3. About 60–70 megs, probably two T3s. It looks like they have a big video stream coming through. Clublove is a dinky site. Cybererotica doesn't even have an autonomous system number. Bandwidth is not indicative of how much money you make. You can make millions of dollars with TI if you're running nothing but text. But once you start doing video and pictures, it eats up more bandwidth. For every dollar made, how much is kept? Cybererotica does lots of webhosting and buying of other people's traffic. He has great gross but shitty net. While Sex.com has tremendous gross and a 98% net, I don't need to buy traffic. That is what separates the men from the boys."

Curious to learn more, Ford drove down across the border to meet the man once and for all. "I wanted to see this infamous Cohen everyone always talks about," as he blogged soon after. Once in Tijuana, he came to a busy thoroughfare with a median of palm trees. Strip malls with UPS stores, tanning salons, and dentist offices lined either side of the road. He pulled into one spot along Diego Rivera Avenue, and went into a brand-new office building, where he was greeted by Jim Powell, Cohen's gray-haired associate of many years. Powell showed Ford around the computer room, which buzzed with servers and wires and heat. It seemed impressive, but, then again, he had no idea who owned what. But the man himself was nowhere to be found, having gone to Vegas, he was told.

Before long, Ford had had enough. He considered the industry leaders his friends, he later wrote, and descending into the belly of a beast, where Cohen and his cronies were suing everyone into oblivion, was making him feel queasy. "These people are into a very heavy revenge trip that goes well beyond scary and it seems to me they do not care about the harm to our industry that they will cause," he later wrote. "I felt so sick after seeing what I saw that I just made an excuse and left Tijuana."

Kremen wanted money. That's what he was going to tell Warshavsky, the online porn wunderkind who, along with Fantasy Man, had offered to fund his lawsuit against Cohen—but had never come through with enough cash. Kremen was heading up to meet him at his office in Seattle. As he drove through the misty streets of downtown Seattle, he felt broke, besieged, anxious, unable to fathom how much longer he could fight Cohen—who was growing increasingly successful.

When he arrived at the headquarters of Warshavsky's Internet Entertainment Group (IEG), it looked like some nerd's porn fantasy come to life. While rows of geeks worked computers in one room, he'd devoted one warehouse to what he called "the Arcade." Inside rooms decorated like dungeons and gyms, male and female strippers cooed into computer screens as the visitors in the tiny little windows eyed them hungrily. It cost $24.95 per month, plus an hourly fee, to have them fulfill their fantasies. Warshavsky had set up cameras in their dressing room, so people could watch them change, and a "pee cam" in the bathroom.

But the boss, as Kremen found, couldn't care less about the distractions. Warshavsky was slight and twitchy, with tight blond curls, and a prominent dimpled chin. Kremen noticed Warshavsky's notorious tic immediately, the way he'd snort a long *honkkkkkk* between sentences. He bounced around like a kid off his Ritalin. And, as he made clear to Kremen, he was under siege. In the past year, he'd made $45 million by luring eyeballs with celebrity scandals. In the time since releasing the Pam and Tommy tape, the twenty-six-year-old multimillionaire had become, as the *Los Angeles Times* put it, "the most infamous pornographer of the Internet Age."

He posted nude photos, taken years before, of conservative talk radio host Laura Schlessinger under the headline "Amateur Slut," alongside compromising shots of Tori Spelling and Keith Richards. More recently, he was getting sued by the Archdiocese of St. Louis for exploiting an upcoming visit by the pope to America by launching a porn site under the domain name PapalVisit.com.

As the suit alleged, Warshavsky had "provided advertising for defendant's adult entertainment websites, hyperlinks to these other websites, and an assortment of 'off-color' stories and jokes regarding the Pope and the Roman Catholic Church."

He had hassles, he told Kremen, and not just the Roman Catholic Church. He was tired of being treated like an outlaw. "Look at Viacom, Time Warner or the contents of Cablevision," as he bemoaned to the *Los Angeles Times*. "All of them derive a huge amount of revenue from adult content. But they don't call Time Warner smut." He'd launched a new site, Online Surgery, which was broadcasting live breast implant and liposuction operations, another for psychics and home loans.

To keep going, he alternated between carbs and proteins. His desk was covered in turkey slices and vitamin supplements. He was going to take IEG public, and was battling reports by others in the adult business that he was swindling them or suing them just to maintain control. And he had no more time left or interest in financing Kremen's own battles. "I'm not going to give you money," he said, as Kremen later recalled. Kremen left empty-handed. With Warshavsky out, Levi decided getting involved was more trouble than it was worth. But he remained available for Kremen if need be down the line. "I kept it professional with Gary," Levi recalled.

But Kremen had no idea how he would afford to continue the battle against Cohen on his own. By the time his thirty-sixth birthday came around, on September 23, 1999, Kremen was licking his wounds. No Match.com, no Sex.com, no payoffs to his countless investments in start-ups across the Valley: an internet wine company, a digital watermark developer, a golf cart marketing group. And now, Cohen was trying to dismiss the case on a technicality—that Kremen, under the auspices of Online Classifieds, never really had the rights to Sex.com in the first place.

It felt like too much, too much work, too much wrong, too much suffering. It was times like this he wanted to just pack up his Honda and drive out to the Anza-Borrego desert for a camping trip and fast. Cleanse his body, cleanse his mind. And so that's what he did, driving off until the buildings became trees, and the apartment lights turned into stars. Two weeks later, on October 7, Kremen's old friend and partner Peng Ong took his content management company, Interwoven, public, raising $53.5 million. Kremen, from a remaining investment of $5,000, had just made $3 million. And he knew exactly how he wanted to spend it: by making Cohen pay.

CHAPTER 8

RANCHO

Not just anyone could get a home in Rancho Santa Fe, the posh enclave of five thousand residents thirty miles north of San Diego. Dubbed "the richest town in America" by the Associated Press, Rancho had been home to some of the country's rich and famous for decades: Howard Hughes, Bing Crosby, Bill Gates. The sprawling mansions dotted the winding roads, shrouded in sweet-smelling eucalyptus and lemon trees.

The town prided itself on a strict "Protective Covenant" that ensured homes had the proper acreage between them and designs that were pleasing to the eye. "To buy a property up here you have to invest a lot of money," as the town's planning director told the AP, "but once you invest a lot of money, you don't have to worry about a McMansion going up next door that's flamingo red."

From the outside, the $3 million, eight-thousand-square-foot estate at 17427 Los Morros Road—with its Spanish-style roof, tennis court, guesthouse, and kidney-shaped, palm-tree-lined pool—met all the community standards. There was even a suit of armor in the foyer. Little did the neighbors know, however, that the masterful mogul of online porn, Stephen Cohen, was living with his new family inside.

Rosey had picked out the home the year before, a bucolic estate where, as she later recalled, she hoped to start "a new life" beyond Mexico with her two daughters and Cohen. She didn't know how much Cohen had paid, or where he got his money. "I never asked," she said. She was just happy to have a peaceful place to raise her girls, with good schools nearby, and enough room for her large, extended family across the border to come and go. This suited Cohen perfectly enough. With its rolling hills and moneyed neighbors, it reminded him of his glory years running The Club in Orange County, just two hours north. Like the Mel Brooks line, Cohen finally felt like it was good to be the king. He deserved this life, this fiefdom, ruling the net, legions of wannabes hoping to be in business with him, while his family enjoyed the fruits of his labor. Despite what his mother had always told him, he had amounted to plenty, he thought, puffing out his chest. While his wife and children spent their days poolside under the palms he would travel down to Tijuana to build his Sex.com empire, beam internet over the border, and keep his rival, Gary Kremen, at bay.

He had reason to be concerned. Kremen was showing no signs of backing down and, if anything, digging in even more against him. They seemed to have one thing in common: they were equally obsessed. Despite the fact that Cohen had control of Sex.com, Kremen was capitalizing on his PR as the inventor of online dating to cash in on the buzz over online porn. All Cohen had to do was go online and search for Kremen's name to find that Kremen was, on November 10, 1999, a featured panelist on "integrity in adult entertainment online" at an internet conference in San Francisco. The event was devoted to "the rise in adult content's popularity and profitability on the Web," and Kremen was billed as "former owner of Sex.com." Who did Kremen think he was?

Cohen wasn't just going to stand by idle. He was going to game Kremen into a corner, use whatever powers he could to drain his time and money. And he had just the plan. Two days after Kremen's adult entertainment panel, Cohen had his lawyer flood the fax machine of Charles Carreon, Kremen's attorney (and a real freak, in Cohen's mind), with a multipage missive. It was Cohen's deposition schedule for the next month—and it was meant to be daunting. They were going to be held in several locations around the world, including two in Israel, three in Moscow, one in Bangkok, another in Athens. Carreon later decried them as what he called "decoy depositions," meant to tie up time and money by sending him and Kremen on a global goose chase. "Never had I heard of anyone abusing the discovery process so blatantly," as Carreon put it.

But Cohen wasn't stopping there. He subpoenaed Kremen's doctor, his old lawyers, a half dozen old business colleagues. It would take about 222 hours and $17,000 in airfare alone to meet Cohen's demands, Carreon told the judge. The judge agreed that it was outrageous, and said as much to Cohen. But Cohen wouldn't stop at that. Instead, he filed a counterclaim against Kremen, alleging that it was Kremen, not him, who had illegally taken control of the Sex.com name, and was now unfairly competing and defaming him—to the tune of $9 million.

Cohen could feel Kremen cowering even more, especially when he learned, in December 1999, that NSI wasn't backing down against Kremen either. Kremen had never had a contract with NSI for Sex.com, never paid for the domain, and, most important, NSI had no way of knowing if Cohen had forged the letter taking over the site. Therefore, they insisted, in a formal motion to the U.S. District Court of the Northern District of California, that they should be dismissed from the case entirely. The court would soon grant the motion, which dealt a devastating blow not only to Kremen but anyone who tried to defend their right to own something online. It determined that a domain name is intangible property, not the kind of property that could be stolen.

NSI, in essence, was telling Kremen to go fuck himself, just as Cohen was telling him. And so, when Cohen found out that he was to appear in a San Diego court reporter's office in February to be deposed by Kremen, he wasn't about to cower. The future of the internet, and his empire, was at stake. He would readily attend—not only to tell his side of the story, but to, once and for all, meet his nemesis Kremen in person and show him the real emperor of love and sex online.

The world did not end at the stroke of midnight on January 1, 2000, despite the warnings. Rumors had burned around the world that a mysterious computer bug, nicknamed Y2K, would cause machines to go haywire at the turn of the millennium. Because most software represented years by their final two digits, the concern was that when the clocks hit "00" the computer would reset its calendars to 1900 instead of 2000—causing programs to fail.

Fear spread that hospitals would go offline, planes would crash, stock markets collapse. A cottage industry popped up around Y2K preparedness, resulting in more than $300 billion being spent in anticipation. When the ball dropped in Times Square, however, so did the worries, it was all for naught. But it seemed to represent something else: how increasingly reliant the world was on computers, and how much the global culture and economy hinged on them.

One month later, Gary Kremen found himself looking out on

the ocean from his hotel room in La Jolla, ready to reclaim his throne online. He was there for Cohen's deposition, and their first face-to-face. He had stayed up all night with Carreon, preparing for battle. Carreon quoted to him from early nineteenth-century Prussian general and military theorist Carl von Clausewitz's classic treatise, *On War*. "In order to induce surrender," as Carreon read, "it is necessary that every day bring the opponent news of a new defeat, and that there be no end in sight to the daily drumbeat of failure."

"Every day has got to be a bad day for Cohen!" Kremen concurred.

On February 3, 2000, Kremen and Carreon drove to a nondescript office building in downtown San Diego, and ascended to the sixth floor to meet their rival once and for all. Flanked by his lawyers, Cohen sat at the conference table. His mouth was in a permanent smirk, like it was chewing his ever-present cigar even when it wasn't there. Dressed in a cheap black suit with a pink and black checkered tie, Cohen eyed Kremen like tobacco he was going to roll and smoke. When he rose to meet him, he handed Kremen a gift: a T-shirt. Kremen unfolded it to see that it bore a logo that read "Sex.Com."

"I thought you'd like this," Cohen said, smugly. "I've sold thousands of them at conventions for $25 a piece."

Before Kremen could respond, the legal teams took their places, and the camera started to roll on Cohen, who smiled enough to show the space between his teeth. He had been in this position before, of course, over decades of cases, and he rose to the occasion, deploying his bag of tricks. When he answered questions, it was with what Carreon and Kremen could only describe as a shit-eating grin, as Cohen responded like Abbot and Costello in their "Who's on First" bit.

"Okay. Mr. Cohen," Carreon said, "you're under oath. You've been under oath before, right?"

"Yes, I've been under oath many times."

"Does the oath constrain you in any way?"

"No."

"You mean you can say whatever you want under oath? Did you understand my question?"

"I understand your question . . . I'm here to tell the truth."

"Right. But for every one that you can recall, you understood the oath to mean the same thing that you understand it to mean today?"

"For every one that I can remember, yes."

"Okay. And which ones can you remember right now?"

"I can't recall."

"You can't recall even the first time that you were under oath?"

"No."

There was a lot Cohen couldn't remember: names of schools he'd attended, his age when he'd graduated high school, what he'd studied during his short stint in college, what businesses he'd had as a young man, when he'd suffered a heart attack that kept him from taking the California State Bar, or why he'd shown up to the deposition essentially blind—because he'd broken his glasses—and was therefore unable to read the documents handed to him. It got more bizarre from there. With a calm, almost patronizing tone, Cohen recounted the most incredible tales as if he was describing the lunch he'd had that day at his favorite restaurant, TGI Fridays. He spoke of being a lawyer for a computer company at one point.

Among his many outrageous claims, perhaps the most outrageous was that he was licensed to practice law in Panama. What was he doing visiting Panama? Carreon wanted to know. "Visiting

Manuel Antonio Noriega," Cohen matter-of-factly replied. Kremen burst out laughing, just picturing the absurd scene of Cohen smoking cigars with the notorious dictator. Cohen was such an intrepid liar, Kremen believed, and an imaginative one at that. But Cohen insisted it was true, saying he had visited Noriega "several times" and had had many business dealings with him, though he wouldn't get into specifics.

All the while, he spoke like he was addressing a child—even when Carreon pressed him on impersonating a lawyer in California, where he had never even applied to the bar. Unshaken, Cohen began bragging about his legal prowess when he was setting up businesses in Mexico, the Virgin Islands, California, and Nevada. Cohen remained inscrutable but steadfast about other claims in his life. But when it came to answering questions about his days running the swingers club in Orange County, his steeliness gave way to sentimentality. As Carreon handed Cohen a copy of an old newsletter he had made for The Club, Cohen's eyes misted over. "It's nostalgic?" Carreon asked.

"Uh-huh," Cohen said, as he read over his newsletter and slipped into the past. He could recall himself in his robe, the disco ball twirling, the Jacuzzi bubbling, the bodies writhing, he'd been so happy, so adored, it was all he ever wanted, a perpetual party of lust and loins and, in his own way, a feeling of love. "Thank you," he told Carreon, blinking out of his reverie. "Can I have this?"

Carreon said he could, but when he asked Cohen about other documents he'd written, Cohen remained cagey. "I've been the author of many documents," he said, including one that was in the Library of Congress. Carreon asked the title. "Excuse my French," Cohen replied, " 'How to Eat Pussy.' "

"That was your—that was the title?"

"That is the title."

"We can get you the catalog number of that if you'd like," Cohen's attorney Dorband offered.

"That's all right," Carreon replied.

But it was the story of how he allegedly obtained Sex.com that remained the most confounding. Cohen maintained that Sex.com had been the name of a discussion board on the French Connection since the 1980s, despite the nomenclature not existing at the time. It had originally been called Sex.communications, he explained, and lectured Carreon like a high school computer teacher. In an old operating system, Cohen explained, "you were allowed to have eight characters in the front separated by a period with three characters at the end. And Sex.com was the name of the particular chat system that certain users went to have general chat."

As Kremen listened, he couldn't believe Cohen's balls—it was not just that he was lying, they thought, but how calmly he spun his web. When Carreon asked if Sex.com was available to everyone on the French Connection, Cohen didn't flinch. "It had limitations," he explained. "We'd only let certain people into the—into that particular area. It was password protected. And the reason for it was, we had a lot of problems back in that day and age. We at one time opened it up to the general boards, and we had a lot of guys that would come onto the general boards and would see these girls and see that they were swingers, and some of the girls got treated like—as though they were prostitutes. And it caused a lot of animosity among the users, and we had to further restrict it as a result."

"You had to further restrict what, Sex.com?"

"The Sex.com portion of it, yeah. And the same with all the other areas outside of Sex.com, like fetish.com, like—oh, we had a nudist area. We had an area—I don't even remember what the nudist area was called." He spoke pedantically about the technical details, how the BBS migrated from a VAX 11750 to an IBM PC. He said he'd discovered that it had been registered by Kremen after doing a WHOIS search on the site, and brushed off Carreon's question as to why he hadn't just called Kremen immediately. "What was the purpose in talking to Gary Kremen?" Cohen asked.

"You knew that Gary Kremen was the system's administrator," Carreon went on. "He was the only person authorized to transfer this domain name."

"Says who?" Cohen snapped. He explained that once he saw that Kremen's old company, Online Classifieds, wasn't registered in California or any other state, he dispatched his old friend, Vito Franco, to pay Sharyn Dimmick, Kremen's former colleague whom Cohen claimed had the power to assign him the rights to Sex.com, a visit. "Find out all the information," Cohen said he'd told Franco. "Find out what it would take for us to get it." Franco, Cohen went on, was an ex-cop in Hawaii, a private eye, and an L.A. movie producer who also worked in construction—most recently in Mexico for Cohen. He'd also set up the live porn feeds for Sex.com. "He ran the live sex, the sucking and fucking that went on in the rooms," Cohen explained. Cohen said he didn't know how Franco had found Dimmick, just that he had driven up to San Francisco to meet her, gotten the letter signed, and paid her $1,000 in exchange.

Carreon pressed Cohen on the strange inconsistencies that followed. Cohen admitted that Franco had typed the letter on Online Classifieds letterhead, but had no explanation as to why Dimmick was unable to type it herself. "You did not think it suspicious that Mr. Franco found it necessary to create the letterhead for the letter to be written on?" Carreon asked.

"No," Cohen said, who didn't think it was suspicious that the president didn't have her own stationery for her company either.

As for Carreon getting in touch with Franco, that wouldn't be possible, Cohen told him. "He's in heaven." He had died the previous October, Cohen went on. They'd last seen each other in a Jacuzzi in Tijuana. The next night, Cohen flew to Vegas for Comdex, and grew angry when Franco had failed to phone him as planned. "I called his house, because I was upset because he didn't call me," Cohen said, "and then I found out that he had passed away." Carreon asked if Franco had managed to put his meeting with Dimmick on tape or in writing before he passed away. "No, he did not," Cohen said, which was such a shame. "He was going to be one of our witnesses."

Cohen arched his back. "Mr. Kremen told the reporter that I identified myself as a trademark attorney and had called him, which was false. Mr. Kremen also told him that I had stolen the name Sex.com, which is absolutely false. Mr. Kremen also told him that I had engaged in criminal acts, specific criminal acts that were false. To make it very, very short, Mr. Kremen and his attorney have engaged—or attorneys, I should say, have engaged in everything from perpetrating a fraud to the court, in my opinion, to simply outrageous conduct, including fabrications of stories about me, making statements, issuing press releases that are just 'fly by the night make them up' stories in order to ruin my reputation in the community, that have substantially hurt the entities that I have worked for, including my reputation."

"Had you ever spoken to Mr. Kremen prior to Mr. Kremen's lawsuit being filed against you?"

"I'm not sure of the answer to that. I received a phone call from somebody that identified themselves as Kremlin."

"Kremen?"

"I'm sorry?"

"Kremen?"

"Yeah. Kremlin."

"Kremen, K-r-e-m-e-n."

"Yeah, Kremlin."

"No. K-r-e-m-e-n. No L."

No matter, Cohen said, the whole conversation only lasted a matter of seconds, and "the person on the other end of the phone sounded very elusive. He just didn't sound like he was all there. And I don't know if I ended the conversation or hung up on him. I probably did."

"Is it your testimony that you did not steal Sex.com from Mr. Kremen?"

"I just testified to that."

"Well, I don't think you answered the question directly."

"I'll answer it again."

"I'll ask it again so it's clear."

"I understand the question."

"I'll ask it again."

"Go ahead."

"Did you steal the domain name Sex.com from Gary Kremen?"

"The answer is no. If anybody stole it, it was Gary Kremen stealing it from me."

Two weeks after the deposition, Cohen sued Carreon and Kremen for libel, and wanted $50 million in damages. But by the summer of 2000, they had a powerful new addition to their team: James Wagstaffe, one of the most powerful lawyers in San Francisco. Growing up in Menlo Park and educated at Stanford, Wagstaffe made his name in high-profile media cases. When Kremen approached him for help, he thought the underlying stakes were compelling enough to get him on board. Compelled by this basic idea, as he put it, "the issue of whether a domain name was like tangible property and therefore you could steal it."

In July, Wagstaffe in tow, they met with Cohen again for his remaining deposition. As Wagstaffe listened skeptically, Cohen painted himself as the international mogul, splitting his time between his homes in Tijuana, Amsterdam, Vegas, and Rancho Santa Fe. His travel was making it difficult to schedule his next deposition. He said he had to go to Europe, though wouldn't say where, to meet a pornographer named Bjorn Jorgenson. But, regardless, he claimed, he wasn't in it for the sex anymore.

"I haven't logged on to Sex.com in a period of time," he demurred. "I'm in the sex business, sir. A lot of this stuff is intriguing to a lot of people that are outside of this industry. I've been in the sex business for many, many years, and some of the things that turn a lot of people on today are just not so important to me in this day and age. You know, I used to run a swing club and I lived a very precarious life. I'm very happily married and live a monogamous life at this point." Still, he bragged, of the five thousand emails he received a day, "there's a certain percentage of those emails where people want to have sex with me."

But then, in the middle of his spiel, he became paranoid. He accused Kremen of making faces at him. "Would you please stop?" he said to Kremen, then to Wagstaffe. "Ask your client not to give me looks when I'm testifying. I know he's got a drug problem and it bothers me." Wagstaffe arched his brow. *What was Cohen talking about?* This just seemed another absurd ploy for reasons he couldn't understand. Though Wagstaffe said he hadn't noticed Kremen mocking Cohen, he said, "I suggest that if Mr. Kremen is making a face or some other action that quietly disturbs Mr. Cohen, that Mr. Cohen focus somewhere else."

And yet, before long, Cohen was at it again. He kept glancing over at Kremen, who seemed to him to be snickering at his every claim. Cohen could feel his anger seething at his rival. Kremen, he thought he was so smart, so savvy, he thought he could just sit there and laugh at him when he was the one who'd be laughing last! "I'm having a problem with your client, you know, smiling," Cohen told Wagstaffe.

"Well," Wagstaffe replied, "let me just for the record indicate that we have been joined by Charles Carreon, one of Mr. Kremen's attorneys. And Mr. Kremen smiled a hello in salutation to Mr. Carreon. That seems appropriate to me. But let's continue."

Cohen barked back at Wagstaffe. "This has been continuous for the last ten minutes!"

"I think that you're—you're not accurately portraying what's going on in the room."

"That's where we agree to disagree."

When Cohen wasn't acting paranoid, he was trying to impress Wagstaffe with his genius at turning Sex.com into one of the biggest empires online. "What you see today is my creation," Cohen said, "and without my abilities to have created that and built that into what it is today—that's what gives it its innate value." The related rights to the site—marketing, television, fiber optics—were worth at least $2 million, he said. Sex.com was getting a whopping forty million hits a day, with nine million people paying $24.99 per month for memberships. He'd said he turned down one offer of $48 million, cash, "because if the company follows my logic and builds this fiber infrastructure, this company will have a product that will be worth between $60 and $100 billion."

As for his checkered past, Cohen stiffened his back and accepted responsibility, martyring himself. "I was engaged in a lot of criminal activity," he said. "I have a very bad past. You are aware of my bad past. There is nothing I can do to make up for my past. If you want to take me out in the wood shack and beat me, that's all I can say. I can only go on, when I came out of jail, I tried to correct my life. And I assume that when I signed this document, I did it honestly, but I cannot represent to you or—and to this court that I was not engaged in criminal activity at the time I did this, because I certainly was."

"When you say you did bad things, you did bad things in business, right?" Carreon asked.

"Sir, as I've already testified, I was a criminal."

"Yeah. And you were a business criminal. That's what I'm trying to say."

"I don't make the distinction. I was a criminal. I committed criminal acts and I was judged by a court to be a criminal. I was a criminal. I'm remorseful about it. There's nothing I can say about it."

"Okay. Well, no, actually I'm sure there's some things you can say.

"You know, what you're telling us, what you're telling the judge and the jury, is that you kind of have undergone a conversion, right?"

"I would like to think so."

"Okay. And that conversion is evidenced by the fact that you live your life differently now. You get up in the morning, you comb your hair, you like yourself because you know you are honest, you're not lying or cheating to get money, right?"

"That is correct."

"You were in the sex business," Carreon concluded, "you ran the swing club and you ran French Connection BBS, and that included services that helped people hook up sexually. Is that it?"

"I had a very active sex life. I'm proud of that."

"And so the question is, you're really still in the sex business."

"I've never gotten out of the sex business."

"So if we were looking for things about your life before prison that were different from your life now, we wouldn't find it in the area of a chosen occupation, right?"

"I don't understand it. Say it again."

"Okay. What I'm trying to do is I'm trying to see—because you've kind of made it important. I'm trying to see the old Steve Cohen and the reformed Steve Cohen, and I kind of have a list of ideas about what the old Steve Cohen was doing, and I'd like to complete my list of what the new Steve Cohen is doing. Now, the old Steve Cohen was in the sex business and the new Steve Cohen is still in the sex business."

"That is correct."

"And, in fact, the business you are in is all based around the continuity of business efforts that started in 1979 with French Connection BBS and continued uninterruptedly until the present day and now is your baby, Sex.com."

"It's my baby, Sex.com."

"So bottom line is your current employment, your current business activity, is just completely continuous all the way back to 1979. You've been in the sex business."

"I'm a man," Cohen replied, "and I've been a man all my life. When I was in the swinging business, I was engaged in parties where I'd fuck and suck 30 women at a time or 20 women at a time. That is not the business I'm in today."

By the time they were done with the deposition, the teams were huddling in the hallway, doing their postmortems. But Carreon kept noticing that Cohen was catching his eye, smiling in a kind of taunting way, like he had something on him or something else up his sleeve. Finally Carreon had enough, and approached the small man in the purple and black tie. "Where are you going for dinner?" Carreon asked him.

"I don't know," Cohen replied, "where are you going?"

Kremen thought having dinner with his rival was a crazy idea, but Carreon convinced him it'd be a prime opportunity for recon. "We don't have to say anything at all, if we're afraid of giving up some advantage," he told him, "we could just listen. If we keep our ears open, we could probably learn some things."

The next thing Kremen knew, he was sitting at a white-tablecloth-covered table in a chop house called Rainwater's, sharing a bottle of red wine with Cohen himself. Cohen kept egging Kremen on throughout the dinner. "You didn't invent online dating," he told him, "I invented it." He went on and on about the French Connection, The Club, his innovations, so much so that Kremen began feeling dizzy from it all. He thought Cohen was a genuine sociopath, someone who believed his own lies, who had a compulsion to impress. He believed he was a lawyer. He believed he invented online dating. He believed he created Sex.com. "You don't understand trademark law," Cohen told him, "you stole it from me."

Kremen searched Cohen's face, the smug smile, the glint in his eye. Jesus, what if this guy was right? And the craziest thing of all: Kremen felt himself beginning to believe it too. "Cohen is a brilliant con man," as he later put it. "He's so convincing that, after twenty minutes of talking to him, he had me believing that I stole it from him." He couldn't help himself. Kremen was starting to lose his sense of reality. Maybe Cohen knew something he didn't. Old insecurities crept back into his mind, like he was getting beat. And Cohen, sensing the vulnerability, tossed him a lifeline. The Sex.com business was going into the toilet, Cohen told him, the real future was in telecommunications, like the internet service he was beaming to Mexico. Cohen told Kremen he could tell he knew what he was talking about, with technology at least. He was an old-timer online, like him. And so Cohen was going to take pity on him and offer him a deal.

He would give Kremen $500,000 to settle this case, in addition to a piece of his telecom company, Omnitech. All Kremen had to do was to let him keep Sex.com and walk away. But Kremen took one look at him and knew exactly what he wanted to do: fight. Fight to the end, because the most valuable website online was rightfully his. And no one, especially not Cohen, was going to take it away. If Cohen wanted a smackdown, he was not only going to lose, he was going to be convicted of stealing from him. "I'm not going to settle," Kremen told him, "and you're going to jail."

CHAPTER 9

THE DOGPATCH

There was the Castro, the Mission, North Beach, Cow Hollow. And then, SoMa, Embarcadero, Telegraph Hill. But while the origins of the nicknames behind San Francisco's many colorful neighborhoods were widely known, the story behind the Dogpatch, the eastward enclave of grimy warehouses and gritty streets, remained a mystery. Some traced it to the *Li'l Abner* days of the Depression, when poor shipyard workers in the area were likened to hounds. Others figured it got its name for the scavenger dogs that would wander the slaughterhouses for scraps.

By 2000, the Dogpatch teemed with homeless people and drug addicts. And Gary Kremen, who felt like his life was going to the dogs, was more than comfortable living that way. While Cohen was enjoying the fruits of his Sex.com millions in his Rancho estate, Kremen had moved into a two-story, run-down old building in the Dogpatch that he bought for $500,000 cash. Kremen had always lived and felt like a pioneer, happy to be alone in the desert or prospecting on the new frontier online.

But still, living in the Dogpatch was extreme. The "war," as

Carreon kept calling it, was taking its toll. Kremen was sleepless and weary, crashing on a mattress on the bare concrete floor, subsisting on takeout burritos and antidepressants. Chasing Cohen had become a full-time job, a 24/7 obsession, plotting all day with Carreon and Wagstaffe, then staying up all night searching the web for anything he could find about Cohen, his money, and his burgeoning empire. Kremen had turned into a kind of DIY detective, unearthing documents that he'd then turn over to Carreon, who'd subpoena them—to put added pressure on not only Cohen but his associates. Cohen wasn't just making money, Kremen realized, he was among the most successful businessmen online. According to a U.S. House of Representatives report, online porn was the most lucrative product online—averaging $2.7 million per day in earnings. Everyone was rushing to make money online, but few were actually showing profit. Cohen seemed to be beating them all.

Soon after, Carreon showed up one morning in his royal blue Jeep Grand Cherokee. They were going to take the deposition of Román Caso, an employee of Cohen's who was working as his vice president of YNATA Corporation, down in Ensenada, Mexico. As the rolling hills passed by, Kremen's cell phone rang. It was Cohen. He knew they were coming and was calling to taunt Kremen personally, to remind him that he was going to jail, that his trip was futile, that he was the original inventor of online dating, and so on. Carreon listened in awe as Kremen took the bait, taunting Cohen back that he was going to be the one relegating him to prison in the end.

As the day turned to night, Kremen's cell phone kept ringing from Cohen. As they settled in their motel room, Carreon on the couch, Kremen kept talking with Cohen, poring over the case, talking about the online porn business, and, to Carreon's amazement, trading ideas. Kremen was probing Cohen about the banner ad business, sharing strategies he'd learned running Match. Cohen would talk about how much smarter he was than the pornographers who populated the underworld. One minute they'd sound like old friends—swapping stories of their Jewish upbringings—and the next threatening to sue the other into oblivion.

After being up all night, Kremen slept through the deposition the next day—waking to find Carreon return, more incensed than ever. The deposition with Caso had been a bust. Carreon had to speak with him in broken Spanish. To make matters worse, Cohen had shown up too, and was insinuating himself in the questioning—cursing at Carreon in Spanish for Caso's benefit. "*Pinche pendejo*," he said to Carreon, "fucking asshole." Fed up, Caso ultimately cut it short, saying in Spanish, "I'm not sure who's the asshole here," then refused to answer more questions, and left.

When Kremen heard the story, there's just one place he wanted to go. So the two of them drove up past the Mexican border and veered off Interstate 5 until the strip malls gave way to rolling hills and luxury homes. They turned along the orange groves, past the horse trails and estates until they came to 17427 Los Morros Road: the address they had for Cohen's home in Rancho Santa Fe. No one was home, but no matter. Kremen just wanted to see it firsthand. Carreon grabbed a disposable camera and climbed into a eucalyptus tree to take pictures for their records. But Kremen just stood there, frozen, soaking it all in. The tennis court, the kidney-shaped pool, the scent of the lemon trees filling the air. Kremen was in the Dogpatch now, but one day, he resolved, this would all be his.

"Gary's coming after you, see?" Jack Brownfield, a silver-haired Texan with leathered skin and sharp blue eyes, told Cohen one afternoon as they sat poolside at the Rancho estate. "He's just going to start taking all your stuff," Brownfield went on, gesturing to the lavish home, "and you've got one big plum hanging out here."

There weren't a lot of people Cohen respected, let alone listened to, but Brownfield, one of his closest business associates and oldest friends, always had his ear. An ex-Marine and Vietnam vet turned lawyer, Brownfield had first met Cohen in 1972, when Cohen was renting office space in the law firm where Brownfield was practicing. Brownfield soon learned of Cohen's business selling bootleg software. His first impression of Cohen never changed. "He's a con man," as Brownfield later put it, and one who seemed always haunted by the specter of his father's disapproval over his schemes. "His father told him he didn't want to ever see him again," Brownfield recalled. "I don't know what his father got upset with him about, but I know it always bothered Steve because he was always trying to show his father that he was doing better financially."

Brownfield had his own criminal past. As a young man, he had gotten mixed up with drug smugglers in Mexico, running coke and speed across the border—only to get arrested and sent to prison for a year. Not long after he got out, he was watching the evening news one night when a report came on about the owner of Sex .com who had offered to buy Caesars Palace, Stephen Cohen. *That son of a bitch,* Brownfield thought with a chuckle. The next day he was at work when the secretary buzzed him. "Hey, you got a phone call," she said, "your lawyer."

It was Cohen. "Hey, were you watching the news?" Cohen said.

"Yeah," Brownfield replied, "what are you doing over there?"

Brownfield had heard decades of schemes from Cohen—The Club, French Connection—and, as Cohen went on about Sex .com, it sounded like his typical approach. "He never has a plan," as Brownfield put it, "he takes off and goes." Sex wasn't the motivation, Brownfield knew, it was money and ego. "He wanted to show people that he was doing well," Brownfield said. "It was very, very important to him."

When Brownfield told him he was skeptical of the online porn business—"look at the scumbags you're dealing with"—Cohen pushed back. Sex was fueling the internet more than anything, people were just too embarrassed to admit it. But anyone with a brain knew where the money was sitting. Google, Yahoo!, America Online, how did Jack think they were making money? By people searching for porn. By driving traffic. By selling ads based on that traffic. And so on and so on and so fucking on. "Jack, this is the business to be in because it never goes out of style!" he told him. "You plant yourself solidly in this business, it's a good business."

Though Cohen had long pretended he was a lawyer, he knew Brownfield was the real thing—and wanted his help with Kremen. When the two got together at Rancho, Brownfield pored over the documents, including the letter Cohen had sent to Network Solutions. Cohen was street-smart, he thought, but bookishly dim. Brownfield shook his head in disbelief. "Steve, this doesn't make any sense," he said, pointing out the typos and inconsistencies. "How come all these letters and everything are all, you know, spelled wrong and everything?"

Despite Cohen's insistence that he was beating Kremen in court, Brownfield suggested he keep as much from Kremen's paws as possible—starting with the house in Rancho Santa Fe. "You need to go and get a loan on this house, Steve," he said. "You get a big mortgage, as big a mortgage on it as you can, so if he takes it, he gets nothing." And just like that, Brownfield was in business with him—securing the loan, and becoming the advisor for his burgeoning empire.

With Cohen now pocketing about a million a month from Sex .com, there was money to go around. Brownfield moved down to Tijuana, renting a place and spending his days with Cohen, who ran his empire from his office, dressed in sweatpants and T-shirts. He helped him with personal matters. Cohen's mother had financial problems in Vegas, and was losing the equity in her house. Cohen had Brownfield give her money every month, collect her bills, and try to analyze how much she was spending and where the money was going.

Cohen was traveling back and forth to Amsterdam, trying to provide porn for the numerous new sites popping up online. Cohen had figured out that the best way to make money online wasn't to create content at all. "He said it's easier to just steal the content from other people," as Brownfield later recalled, "and edit it so it looks like original content, and then reproduce it that way rather than trying to hire the people to do the directing."

Back in Mexico, Cohen was busy building his fortune there. Brownfield drove with Cohen across the border to see the small shack with the microwave dishes. Cohen, cigar between his teeth, described how he was building the biggest telecom in Tijuana. Brownfield didn't pretend to be technically savvy, but he knew enough to see red flags. Cohen's system seemed like basically a big hack, allowing customers to make international calls and get online service outside the regulatory system.

Before long, Brownfield grew weary of giving Cohen so much advice without getting something more substantial in return. "You know, you've got to give me something," he told Cohen, and he had just the thing in mind: a shrimp farm. Brownfield had already been studying aquaculture, and researching the business in Mexico. He'd spent time studying shrimp farming in Culiacán, a city in northwestern Mexico, and getting licensed. The fishing industry in the country had long been nationalized, Brownfield explained, but with NAFTA this was no longer the case. And there was big money to be had.

"The differential between what you can buy the shrimp for in Mexico and what you can sell it for in the United States is over a dollar a pound," he explained. "You could ship everything that you could load on a plane you could make a dollar on. If you're selling five tons a day, you're making $5,000 a day, right?" Drawing from his old drug smuggling days, he devised a way to ship the shrimp by plane instead of the usual boats—so they could increase deliveries to make even more cash. And with shrimp consistently outselling other seafood, investing there would be a no-brainer.

Cohen was dubious. "Why do you want to do that, man? We could do anything."

"No, Steve," Brownfield replied, "this is what I want to do." He didn't want to make money from sex or drugs or anything outlaw anymore, he wanted the *camarones*. And so, Cohen agreed to invest in one for Brownfield. But in the meantime, there was the Kremen problem, and Cohen had just the plan for making him go away.

His name was Crab. Just Crab, a human crustacean, a wily, clawed scavenger of the Dogpatch who latched himself to Kremen's side. Kremen's industrial home had assumed the nickname of his grimy neighborhood, and street urchins such as Crab were coming with the territory, insinuating themselves inside the walls of his house and mind. It had begun, at some point, when he let a homeless guy start passing out on his couch. Before long, there were others coming and going, rifling through his fridge and shooting the shit while he hunched at his PC, trying to dig up what he could on Cohen. Though he had way more education and money than they'd ever know, he identified more with them than the MBAs and dot-com wannabes he left behind at Stanford. They were survivors, salt of the earth, scrappy pariahs clawing for scraps just like the crab that he too had become. And when Crab, the ruler of them all, offered Kremen a pipe late one night, he didn't hesitate to take a long drag of what was inside: crystal methamphetamine.

Kremen's history with drugs, until that point, had been fairly run-of-the-mill for someone of his socioeconomic class and culture. Never much of a drinker, there'd been weed in high school, acid and mushrooms at Dead concerts, cocaine now and then, and that was pretty much it. But meth just grabbed him on a lark out of nowhere like nothing he'd tried before. From the moment the sweet sizzling burn hit the back of his throat, he was up again, a mad rush of dopamine and energy and confidence mainlined to his brain, a fat pipe of bandwidth streaming in the necessary connection to keep him awake and alive and fucking focused, so focused he could do anything, could stay at his PC for fucking hours, Grateful Dead on the stereo, Crab laughing about something, a hot burrito in his paw, and then, when he felt himself dipping, another hit of the pipe.

Kremen always heard the sound of his mother's voice in his head, telling him to do better, work harder, beat Cohen. But that voice took on a darker obsession now. He needed more energy, more fuel, and he found it in meth. When Carreon showed up to work, he could tell that something was amiss. Kremen was always a nervous guy, but seemed increasingly so now—hair matted down with sweat, eyes shifting. But then again, he was a grown man, he could make his own choices, and seemed, nevertheless, functional despite whatever he was consuming. And, if anything, he was turning up even more on Cohen—financial documents, bank accounts, associates. Carreon didn't know how or where Kremen found it all, but had a strong feeling that Kremen was not only a deft researcher on a computer, but someone capable of hacking into places he wasn't intended to go.

By the fall of 2000, Kremen was still investing in other start-ups, hoping they'd cash in. But his full-time passion was getting back Sex.com. Between his sleuthing and his lawyers' subpoenas, served with the help of a network of private eyes, they were getting a clearer picture of just how much money Cohen was making, and where he was stashing his cash. This wasn't just about winning Sex.com, after all, this was also about Kremen getting the cash that Cohen had made from the stolen site. Every clue they found showed them what money was to be had: the house in Rancho Santa Fe, his bank accounts, his securities accounts, Citibank, Charles Schwab, his corporations, Sandman, Ocean Front. They had tax returns, stock certificates, letters from the French Connection to Wanaleiya, papers in and papers out, bank documents in, bank documents out. Kremen's desk in the Dogpatch grew with stacks by the day, as he hunched over them with his red pen, circling points, and scribbling in the margins.

In September 2000, they were waiting for details on Cohen's essential bank accounts at Wells Fargo, which they had finally tracked down. They were due to be sent to a Kinko's copy shop in Chula Vista, California, they had been using to process the subpoenas. But when one of Carreon's assistants called the Kinko's to see if the paperwork had come, she received a surprising reply: the Kinko's manager told her that one of Kremen's lawyers had already picked up the documents. When she asked for a description of the lawyer, the manager told her he was a stout, short, graying man with a receding hairline.

Kremen was in Wagstaffe's San Francisco office when he got the call from Carreon: "Cohen stole the documents!" he said.

Why he would go to such lengths, they had no idea. After all, he could have simply requested his own files from the bank at any time. Weirdly, soon after, a FedEx arrived at their offices without a return address. Inside were Cohen's Wells Fargo records—with two hundred pages missing. It seemed Cohen must have wanted to intercept the documents, remove some key pages, they assumed, then mail them along as if they had come straight from the bank.

All they needed was proof. And, soon enough, they had it in surveillance footage from Kinko's that showed, in grainy black and white, Cohen, in an oversized sweatshirt, coming to the counter, and leaving with the stack of bank documents. A few moments later, he came back into the store, grabbed a large FedEx envelope—the same one they now held in their hands.

Carreon viewed the moment in typically philosophical terms. As he later wrote, "Sun Tzu compares a force attacking with momentum to a raging flood tossing boulders like pebbles. A good way to generate a raging flood is to build a dam, then break it." After months of slogging through the courts, the videotape of Cohen's theft was just what they needed to break the dam. And there was no time to waste. Every month, Carreon discovered, Cohen was getting his Sex.com profits sent to the account of a company called Omnitech—as much as $1 million.

In total, they found, Cohen had grossed hundreds of millions, clearing $43 million from ads and membership fees, more than half of what the online auction site eBay had earned in its entire history. Sex was not only the biggest source of gold online, Cohen was at the end of the rainbow with the biggest pot. But there was a twist. Every month, Cohen was draining the account by transferring the funds to banks in Luxembourg. There was no doubt about it. He was laundering his Sex.com cash, and hiding it offshore. Pam

Urueta, another attorney on Kremen's team, likened the hunt for Cohen and his money to "trying to nail a jellyfish to a wall."

On November 27, 2000, it was time at last to find out if Judge James Ware, who was presiding over the case for the U.S. District Court for the Northern District of California in San Jose, agreed. Kremen and his team filed into the courtroom, only to find that Cohen, not surprisingly, was nowhere in sight. As Ware listened intently, Wagstaffe firmly recounted the undisputed evidence: Cohen's forgery of the letter to Network Solutions, his subsequent hijacking of Sex.com, his impersonation of a lawyer to steal the bank documents from Kinko's. And, despite the best efforts of Cohen's attorneys that day, Judge Ware agreed. "Cohen's interest in this domain name Sex.com is based upon what appears to the court to indisputably be a bogus document," he said, "and the consequence of that is to take the name back to its original registrant—and that is Mr. Kremen."

And, just like that, it was done. Ware ordered that Sex.com be returned to Kremen once and for all. Furthermore, because Cohen, as Ware wrote in his judgment, had "engaged in activities designed to conceal money" by moving his Sex.com profits offshore, he ordered Cohen to provide a full accounting of every penny he had made from the site. In the meantime, until the final damages were determined, the judge ordered him to put $25 million into a federal court account.

When Kremen heard the news, he felt a rush of dopamine like a thousand hits of meth. Outside the courthouse, the press descended upon him, as he pumped his fist into the air, relishing the victory. The win was considered a pivotal moment in the evolution of the new world online. Reuters wrote that it "marked an important step in resolving the issues of Internet claim-jumping and the real value of cyber real estate." The *San Jose Mercury News* called it "a significant victory for Kremen in one of the most widely watched domain-name disputes in cyberspace." As Pam Ureta put it, "This ruling sends out the clear message that the domain name is a valuable property right that can and should be protected."

As Kremen gazed euphorically upon the crowd, he could feel his life changing. After spending countless days and nights, as well as $2 million in legal fees, he had proven victorious. All the pain, all the suffering, the late nights, the sleuthing, the nightmares, done. "It shows that eventually the little guy can win at a great cost," he told a reporter from *Wired*. When asked what he was going to do with the site now that it was his, Kremen said he was going to clean it up for good. He was thinking of making it like an educational Google for sexuality and relationship links. He still thought of himself as a nice Jewish kid from Skokie, Harriett and Norman's son. What Cohen had done had "no value," he said. "I plan to do something not as disgusting and sicko as this guy."

But no matter what he did, one thing was true: he was now the owner of the "hottest address on the web," as one reporter put it, with an estimated value of $100 million. The inventor of online dating was, for the moment at least, the accidental king of online porn. And he had three missions: to build Sex.com into his own empire, to get an apology—and perhaps financial compensation—from Network Solutions, and to wring from Cohen every single one of the tens of millions of dollars that his nemesis had already made from the site. He may have woken up in the Dogpatch but he wasn't going to sleep with the hounds anymore.

CHAPTER 10

THE CRAZY AUNT IN THE ATTIC

On March 10, 2000, investors on Wall Street stumbled over each other to buy rounds of drinks. They were celebrating the so-called new economy coming from Silicon Valley. That day, the NASDAQ hit its sixteenth record number of the year. Buoyed by technology giants such as Dell, Microsoft, and Qualcomm and day traders investing from their computers at home, the market broke above 5,000 points—capping off one of the business world's most incredible runs in recent memory, increasing five-fold from 1995. The combined value of tech stocks was $6.71 trillion.

The internet, once the purview of nerds and pariahs like Kremen and Cohen, had boomed into the mainstream beyond anyone's imagination. The number of people online had grown from 16 million to more than 300 million in five years. In 1995, only 14 percent of Americans used the internet, but now the number was 46 percent. As a result, people were spending more and more time online—over four hours a week—and leaving old media for new media. More families in the United States now subscribed to online services than newspapers. Consumers spent $20 billion online the previous year, with Amazon attracting more than 17 million people a month. In January, Steve Case, the founder of America Online, announced plans to acquire Time Warner for $182 billion in stock and debt.

With money to spend, dot-coms took over 20 percent of the Super Bowl ads that month, transforming, among other things, the sock puppet mascot for Pets.com into a national meme. Seemingly overnight, dot-com start-ups such as Monster.com and Kozmo.com, a home delivery service, were the new wunderkinds of the business world. Their motto was "get big fast," offering their goods and services for free to build market share and cash in later. "When you break through 5,000 everyone wants to be in the game," as the head of one NASDAQ trading company told CNN at the time.

At least, it turned out, until the next day. On March 11, the hangover from the party set in—along with a devastating slide. Investors grew scared whether the companies would deliver. By the end of March, the NASDAQ was down to $100 billion. And an antitrust ruling against Microsoft only made this worse. By early April, the market had lost more than $1 trillion. One J.P. Morgan analyst told *Time* magazine that many companies were losing between $10 and $30 million a quarter. "The reality is that many of these companies are simply running out of cash," the analyst said. By year's end, he went on, "a majority of the owners will be forced to turn out their lights and go home."

But not everyone in the dot-com world was living in the dark. The underworld of the web, in fact, was booming. While sites such as Amazon had yet to turn a profit, porn was now generating $1 billion in annual sales internationally. The technological innovations were fueling a boom in cable and DirecTV distribution as well, bringing the total industry of adult content to $10 billion.

Adult sites were pulling in a whopping 35 percent of all the unique visitors online around the world.

But it wasn't just the Stephen Cohens of the world who were cashing in anymore. The biggest corporations in the country—AT&T, Time Warner, General Motors, EchoStar, Liberty Media, Marriott International, Hilton, On Command, LodgeNet Entertainment, or News Corporation—were making a killing. But, as one AT&T executive told *The New York Times,* the last thing they would do was talk about it publicly. "How can we?" the executive said, on the condition of anonymity. "It's the crazy aunt in the attic. Everyone knows she's there, but you can't say anything about it."

If there was one thing people were still buying online, it was porn. More people visited porn sites than government or sports pages. Roughly one in four regular internet users, or 21 million Americans, cruised some of the sixty thousand porn sites online. According to a Nielsen study, porn sites were getting fifty million hits a day. With this kind of success, many of the biggest online porn providers were in business with the largest telecoms around. New Frontier, a publicly traded company with thousands of porn sites, provided sex videos to the largest cable and satellite companies.

AT&T had its Hot Network porn channel—subscribed to by 20 percent of its customers—along with a separate company that sold porn videos to more than a million hotel rooms. As the *Times* pointed out, the General Motors Corporation was selling more porn videos than Larry Flynt, with 8.7 million Americans spending more than $200 million a year in pay-per-view porn on DirecTV, a GM subsidiary. EchoStar Communications Corporation, the second biggest satellite provider, backed largely by magnate Rupert Murdoch, outperformed *Playboy* for sexual content. "Despite the fact that this material isn't marketed, revenue-wise, it's one of our biggest moneymakers," said Peggy Simons of TCI Cable, owned by AT&T. Her comment came, however, only during a court hearing.

For the adult industry, it was only the beginning. "These companies like AT&T, they're thinking ahead to a time, perhaps in 10 years, when 50 million Americans will have broadband capability and all their television and Internet will be interactive through one big box," Bryn Pryor, technology editor for *Adult Video News*, an industry trade magazine, told the *Times*. "But it's not just technology that made the big boys get into it," he went on. "This just happens to be a business where you can't lose money."

On November 27, 2000, the morning of the hearing that awarded Sex.com to Kremen, a stout, balding man walked into the cool lobby of the Wells Fargo bank in Anaheim Hills, and transferred $1.2 million out of his bank account. Cohen wasn't taking any chances. If Kremen was going to win the domain back, then the last thing he was going to do was make it easy for him to get any of his money. As Cohen had told Brownfield, "I'm never going to give that son of a bitch a dime. He didn't earn this thing. I have built this business."

But Cohen's protests were to no avail in the weeks following the awarding of Sex.com to Kremen. The business was no longer his. It was in Kremen's hands, and Cohen went into panic mode, knowing that the rest of his empire—everything he built—was sure to follow. Cohen filed an appeal, as expected, in federal court, and refused to pay the $25 million he'd been ordered to put up until damages were assessed. Wagstaffe pushed for the accounting of his Sex.com earnings. "I have a simple question," said Wagstaffe, suggesting that Cohen was hiding his cash. "Where is the money?"

"We're not hiding anything," Cohen's attorney, Bob Dorband, said. "The defendants do not have $25 million to deposit into the court."

As the lawyers battled, Brownfield tried to reason with Cohen. Instead of fighting Kremen, why not find some way to collaborate. Sure, they had their differences, but they also had a lot in common, Brownfield thought: a love of money, a knack for business, plus traits that complemented each other. "You know what? You two guys need to get your heads together," he said. "Steve, you've got the knowledge, but he's got the common sense to run a business because he's got the education. I mean, here's a developing business, you guys were the first ones in! Together, you guys could just own this industry and sell it, you know, or do whatever you want to do with it. You can *own* this business."

For a moment, Brownfield caught a glimmer in Cohen's eyes, like he was actually considering it. But the moment passed. "I can't work with that son of a bitch," Cohen said. And here was the even crazier thing, Brownfield thought: they were still talking to each other, Cohen and Kremen, almost every day on the phone. Cohen would call him at all hours of the day and night, taunting him, prodding him, while Kremen just sat there shooting the shit. "I'm going to visit my money over in Luxembourg," Cohen might say, only to hear Kremen flush a toilet in the background in response, which would only infuriate Cohen more.

"You don't know anything about this business!" he'd say. "You want to be in the sex business and you don't know anything about it." Then he'd slam down the phone. "He'd hang up and you could see in his demeanor that he was upset or he was angry," Brownfield later recalled. "Steve didn't want to lose that business."

So instead of giving in, he did the same thing that Kremen had done all along: he dug in his heels, and schemed how he could fight back. First, there was his money, the millions, which he had moved offshore. But then there was the gem: his house, the home for Rosey and the kids, in Rancho Santa Fe. It was more than a house, it was a home, it represented everything his parents said he would never amount to. And if Kremen was coming for it, then he was going to have to make sure it would never fall into his hands—one way or another. So he called up Brownfield, and told him there was something he needed him to do.

When CNN asked Kremen his plan now that he owned the biggest porn site online, he shrugged his shoulders. "I still need to figure out exactly what's going on with it," he said. But there was one thing he was sure about, though: "I don't really want it to be a porno site."

As impractical as it sounded to the rest of the industry, it was true. The man with the biggest name in the biggest industry online wanted to be known as Mr. Clean. He made this clear during his first couple months after winning back the site, as he made the rounds among the biggest players in the business. He traveled North America with his self-described consigliere, Carreon, getting the lay of the online land of laying.

They started with Yishai Hibari, the young Israeli ad man who'd helped Cohen build Sex.com into an empire. In exchange for a 15 percent cut of his ad sales, Hibari agreed to work with Kremen, and the two met, along with Carreon, one afternoon at Hibari's new offices in midtown Manhattan. Bald, confident, and stylish, in a slick black suit, Hibari held court in his conference room, flanked by two assistants who, he pointed out, had formerly worked in Israeli intelligence, the Mossad.

On a large screen, Hibari rattled off numbers as banner ads

scrolled down before their eyes. Hibari had been in the business long enough that the images of money shots and open mouths didn't seem to register; it could have just been kitchen cabinets and lawnmowers on-screen. Cohen had a certain genius, Hibari went on: keep the site design simple, keep it explicit, and keep it profitable. The homepage had now been winnowed down to six high-priced ads going for $50,000 a pop, alongside a row of links—to sites such as Big Beautiful Women and Cheerleader Gang Bangs—for around $7,500 each.

Because Sex.com was such prime property, it attracted not only huge waves of newbies looking for porn, but ones naive enough to subscribe to sites—even when they could likely find free porn themselves if they knew where to look. According to the stats, the visitors were clicking on the ads at an astonishing rate: more than two hundred clicks per second, and for each site. In the online ad business, getting just one of these people clicking an ad to become a member was considered a success. Sex.com was averaging four times that rate. Kremen would get a percentage of every subscription, just as Cohen had all along. Even better, Sex.com already had the biggest whales in the business, such as Levi and Warshavsky, among their clients. The money came in wire transfers, paid in full in advance for the ads—in Cohen's case, making upward of $1 million a month, or more. And with estimates of online porn projected to become a $4 billion industry, Kremen's profits were about to explode. Cohen had, in other words, set the table; now all Kremen had to do was sit back and cash in. As one pornmaster told them, "Say whatever the fuck you want about Steve Cohen, he knew how to make money."

But as Kremen took it all in, he could feel his head throbbing—and not just from the gin and tonics the night before. He didn't want to be in the business of sex, he just wanted to be in business.

It felt like the hangover of his battle with Cohen was finally catching up with him. All these months, he'd been so obsessed with winning that he hadn't fully considered what he'd do if he actually won. And now, here he was, dealing in online porn, meeting with ex-Mossad, as the most hard-core images he could imagine stared down from the screen before him. Sure, tons of young guys would kill to be in his position, but so what, he wasn't them. All he could ask himself was: What would his mother think?

And so, his answer to Hibari was simple and swift. He would keep the same business model as Cohen, but he wanted to clean up the site—effective immediately. Within thirty days, he told Hibari, he would no longer accept any banner ads that showed actual sex. He also wanted to strip any references to anything potentially against the law, such as bestiality or incest. And any models on the site had to be clearly of legal age.

After the meeting, Hibari pulled Carreon aside. Kremen seemed well-intentioned but, as another porn sales guy put it after hearing his plan to sanitize the site, "get real, this is pornography." Kremen would destroy the empire that Cohen had so deftly built. Why not get out of the business before he got in? Hibari told Carreon he knew a rich guy who was loaded, and about to buy a phone company in the Midwest from his porn profits. They would buy Sex.com right now for $15 million cash. But when Carreon ran the offer by Kremen, Kremen scoffed. No matter how squeamish he felt about the porn industry, he hadn't worked this hard to get Sex.com back just to give it away—no matter what the price.

From New York, they went to Toronto, to meet the moguls behind two of the other biggest sex and dating sites, Date.com and Orgasm.com (which boasted the slogan, "keeping Kleenex in business"). Kremen showed up in a stained sweatshirt and loose jeans, a thin goatee on his face; while Carreon sat beside him like a roadie: long-haired, blue jeans, motorcycle jacket.

Carreon listened appreciatively as the porn guys paid homage to Kremen's history with Match.com, and tried to feel out his intentions now. Was he *really* going to clean up the biggest online porn site? As a test of his mettle, they took him to the most expensive strip club in town. Carreon blanched, and spent the better part of the night talking with a dancer about her kid, while Kremen embarrassingly endured a few lap dances, as he made awkward conversation with the other guys. But the music was loud and Kremen's body felt like it needed the respite of sleep, a break from the fast track he now felt himself on.

The fact was, he was also jonesing—for cocaine, for meth, for something to keep him up. The drugs had become part of his lifestyle now, a functional lifestyle, nonetheless. He didn't have time to hate himself for letting this happen, he was too busy, too focused, and if anything the drugs were just fuel for the climb. It was only at night, when he collapsed on his bed, that his world would go dark and quiet, between fitful turns of sleep.

Next thing he knew, he was at a kinky Christmas party in Los Angeles, crowded with throngs of well-dressed guys and slinky women gathered around a dance floor as a shirtless young white guy beat conga drums in the middle. The man was one of the most successful veterans of online porn, Gregory Dumas, head of the company hosting this party, New Frontier. Dumas had cut his teeth launching Hustler.com for Larry Flynt, and had gone on to have one of the biggest sites, iGallery, which he sold to a public company for $36 million. Kremen considered him something of a peer in the domain business, since Dumas had gobbled up more than 1,500 sexually related domains. Plus Dumas was funneling content to the burgeoning porn channels in hotels and on cable TV.

But Kremen relished that he had the ultimate dot-com in Sex .com, and even Dumas—a multimillionaire playboy who liked to yacht and fish and party on the drums surrounded by beautiful women—wanted what he had. Kremen that night played it up like the life of the party he never was as a kid. "Where's my escort?" he'd joke with Dumas's crew. "Isn't this a sex company? Aren't we in L.A., ground zero for sex and money? Isn't it Christmas? Don't I own Sex.Com? Where are my presents?" Kremen was kidding, but Dumas wasn't—and offered him $24 million to buy Sex.com. But again Kremen said no. It was enough for him that they wanted what he had, but he wasn't going to sell out, let alone become one of these guys—building an empire on hardcore sex.

In January 2001, their travels took them to the belly of the beast—Las Vegas. The occasion was Internext, the annual trade show held on the heels of the Consumer Electronics Show, for the purveyors of online porn. From the moment Kremen strolled past onion-domed minarets at the Sands Expo, he could see that, despite the dot-com crash still reverberating in Silicon Valley, business was thriving among the moguls here. The year before, the convention was still small enough to be held in a modest conference room with folding tables and chairs.

But with business thriving, they had moved to the largest hall available. Kremen walked in to find more than 170 companies there and some 5,000 people milling around the 150,000-square-foot venue. Booths teemed with scantily clad porn stars teetering in Manolo Blahnik heels; the names of the companies stretched on banners across the top: Hardcore Money, Silver Cash, Babe Net, and more. They were giving away T-shirts—"It's all about blowjobs" and "I Heart Porn"—calendars, giant-size bottles of Astroglide lube from the gay site Outster, stickers from the Free Speech Coalition, and PAW (Protecting Adult Welfare), a social service agency for retired porn stars.

It wasn't just porn moguls making the rounds, it was mainstream business executives looking for the next big thing in online technology. "Pornography is something that is an impulse buy. Technology is a way to help people fill their impulses," Anthony Edmond, owner and CEO of adult internet company Flying Crocodile, told the *Las Vegas Review-Journal*. It was, in other words, a fulfillment of Maslow's Hierarchy of Needs, just as Kremen had long believed. "CNN could not have cnn.com . . . with the old technology," as Edmond went on. "They need our technology."

One such innovation was in the booth for Voyeur Dorm, a reality porn site, filmed in Tampa, Florida, that kept twenty-four-hour cameras on a home of eight women. Sixty-five thousand members were paying $34 a month to watch the women hang out, smoke, shower, and have sex. The one-named owner, Hammil, compared it to emerging reality TV shows such as *Real World*, and he suggested that one day people would be used to watching people hanging around their homes like this. "We've always come up with the next best thing," said Hammil.

Trying to fit in as best they could, Kremen and Carreon navigated the late-night parties. A sing-along in a suite hosted by gay porn director Chi Chi LaRue; male strippers in G-strings and free sausages as hors d'oeuvres. In a ballroom, Ice-T and George Clinton performing onstage as attendees writhed on the floor. Before long, he and Carreon were summoned to the palatial suite of Ron "Fantasy Man" Levi, the godfather of online porn. Fantasy Man, dressed in Johnny Cash black and chain-smoking Marlboro Lights, sketched out an offer he hoped Kremen wouldn't refuse: $400,000 a month to license the Sex.com name, along with 40 percent of the earnings. Kremen said he'd have to think about it.

As Kremen made his way back out, he felt dizzy. Dizzy from hearing the pitches, posing with the porn stars, and inside he began to feel sick. Sick from withdrawal, sick from the attention, sick from the sycophants pulling at his sleeves. He and Carreon stopped at a bar with porn stars in silk pajamas, beers being served out of silicone sex dolls' orifices. But Kremen stealthily slipped away. By the time Carreon made his way back to their suite, he found Kremen sleeping inside, wheezing and sweaty. Standing there looking at Kremen, Carreon got an ominous feeling—that their days together were numbered.

He was right. Before long, their relationship deteriorated into arguments and accusations. Carreon accused Kremen of stiffing him on the 15 percent he said he'd give him of the Sex.com profits, and Kremen pushed back alleging that Carreon had to share in the millions of dollars it took to win the lawsuit. Kremen had no time or patience for Carreon anymore. He was too consumed by Cohen to care about Carreon. The last four words he said to Carreon were "See you in court!"

One day in February 2001, Rosey Cohen, Stephen's wife, was at home in Rancho Santa Fe when her phone rang. It was her husband. In Spanish, he told her to grab the kids and their belongings, it was time to move out. Rosey was used to never asking questions about his business, but knew something was afoot. She had no time to react. In a blur, a group of Cohen's workers from Mexico came to the house, ripping out the seats of the movie theater Cohen had built, and taking them along with the home theater system back to their boss in Mexico. And then Jack Brownfield, Cohen's right hand man, showed up, with compassionate eyes, to help her and the girls move back across the border. And as for Kremen, Cohen wasn't going to pay the $25 million or provide the accounting of Sex.com. He could go fuck himself. Cohen was staying put in Tijuana, he resolved, just let him come and find me.

When he got an order to come up to San Jose for a hearing at the end of the month, he wasn't going to make that either. Instead, his lawyers delivered a message to the court. Cohen would be unable to make the hearing, because he was under house arrest in Mexico for unspecified charges. Judge Ware, however, called the claim "unsubstantiated" and disregarded it.

Instead, he ordered "a warrant directing the United States Marshal's Service to arrest Stephen Michael Cohen and hold him in custody until either the conclusion of the trial of this action or until he performs the following acts: return to the United States and deposit with the Court $25,000,000 or such lesser sum as he shows is warranted by his economic circumstances, return to the United States all revenue generated from Sex.com, provide a full accounting of the Sex.com domain name operation, and effect the turnover of $1.1 million in bank funds to foreign accounts after being ordered by this Court not to transfer any such funds."

When asked about the warrant, Cohen's attorney, Bob Dorband, struck a philosophical note. "It's a very strange case," he told CNN. "It has some unusual characters, who really are more alike than they are different. I think if they had met each other in some different forum they would actually be friends." As the trial date approached, the big question was whether Cohen would show. "I don't know if he will," Kremen told a reporter for *Wired*. "I'm curious." But it wasn't just Cohen who was a wild card, it turned out, it was Kremen's former friend and attorney, Carreon. On the morning of March 8, Carreon filed charges against Kremen, claiming he had breached their contract by failing to give him his 15 percent stake in the Sex.com profits.

But as Kremen took the elevator with his lawyers to the fourth-floor courtroom of U.S. District Court in San Jose, he had other matters on his mind: what, if anything, he'd walk away with this day. As he anticipated, his rival, Cohen, was a no-show. "Mr. Cohen in Oct. 1995 fraudulently induced the Internet Registrar, Network Solutions, to change the registration of the domain name to his own name, and that since that time he has made millions from the operation of the websites," said Tim Fox, a spokesman for Jim Wagstaffe's law firm, which was representing Kremen. "We are asking the court, which has already given us back in November the right to the domain name, to also give us all the profit that Mr. Cohen has wrongfully obtained over the last five years." The amount they wanted: $43 million.

The next month, the final judgment came. Cohen was ordered to pay Kremen $65 million, including a $40 million judgment and $25 million in punitive damages and the house in Rancho Santa Fe. It was, as the *Los Angeles Times* put it, "a record verdict" for the internet and the value of property online. For Wagstaffe, the verdict marked a significant turning point in the evolution of the new world online. "The substantial size of this damage award sends a message that the Internet is not a lawless wasteland," he said.

As for Kremen, the $65 million verdict felt weirdly bittersweet. On one hand, it was a tremendous victory, of course, further validation that he had been right all along, and that Cohen had swindled him. And, yes, the dollar amount was truly staggering—more than he could even conceive—enough to let him live the life of his wildest dreams, and leave plenty for his friends, his family, and any charities he wanted to support.

But within milliseconds, as he was wont, any sense of triumph turned to skepticism. It was one thing to win the judgment, and another altogether to collect it. If there was any way he was going to see a penny of the $65 million owed to him by Cohen—now a fugitive, with an outstanding warrant, hiding in Mexico—it would be the fight of his life. But that was just what he was prepared to do. "I expect to get my hands on nothing," he told the *Los Angeles Times*, but "this isn't about business anymore. It's the principle."

CHAPTER 11

THE PLAYERS BALL

They called him El Sapo, the Toad. Gustavo Cortez Carbajal was a fifty-three-year-old Mexican attorney with dark hair and thin black mustache, and a passing resemblance to his amphibious namesake. He'd been practicing law for more than twenty years and worked out of a darkened office with faded furniture in Tijuana. And that's where he'd sit across from his most demanding client, Stephen Cohen.

With Kremen closing in on him, Cohen was doing everything, and anything, he could to keep his money safe and hidden. There were the offshore accounts, the shrimp farm, the home in Tijuana. And now the Toad had another investment for Cohen—one that was right up his alley. They spoke in Spanish to each other, a language Cohen had mastered. "*El Bolero,*" the Toad told him. "*Vámonos.*"

They went down to the commercial center of TJ, past the sex bars, farmacias, the street vendors selling grilled corn and tortillas. They came to Señor Frog's, a raucous bar, popular with the locals, and then there it was, right next to it: the Bolero, the premier strip club in town. It was a small orange building with a red tile roof on a busy street in the town's red-light district. This was no run-down shack offering donkey shows, as Cohen could see. *Los Angeles* magazine described it as "the Taj Mahal of Tijuana clubs," popular with Hollywood agents, Silicon Valley moguls, and other moneyed tourists looking for sin and business across the border.

There were Ferraris out front, not pickups. As they passed through the doors, the air was filled with the thick smoke of Cuban cigars. Mexican disco pounded from the speakers. Two long bars ran down the middle, peopled by men in suits more interested in making deals than taking any of the dozens of voluptuous dancers on the large, mirrored main stage. Others, however, couldn't resist the temptation, reaching up to take the slender hand of a raven-haired dancer in lingerie, and following her behind a velvet curtain for a $40 single-song private dance, or perhaps more. "Hundreds can disappear in short order," as *Los Angeles* put it.

Cohen liked what he saw: the women, the cigars, the money. What exactly did the Toad have in mind? *Una inversión*, an investment to become a part-owner of the Bolero. A club, Cohen thought, a bit like The Club, his old swingers club in Orange County. To be the man again, the king of a scene like this, greeted by strippers with open arms and open legs. *Por qué no?*

Cohen met with Héctor Manuel Huerta Garza, one of the owners, to strike a deal. Huerta eyed the doughy balding American skeptically, as Cohen began boasting about his fortune. "He told me he was the richest man in Tijuana," Huerta later recalled, "that he had more than a hundred million dollars." He bragged about his internet company in Europe, his home on Playas de Tijuana, the way he'd lend money at a rate of 3 to 1, that he owned his penthouse on Sonora Street, his $2 million shrimp farm.

Cohen told him he had enough to buy all of Pueblo Amigo, the area where the Bolero and other clubs were located. Huerta's eyelids felt heavy the more the gringo spoke. The man struck him as a "scoundrel," as Huerta later put it, but the gringo had money. Their deal was $800,000 in exchange for 40 percent of the Bolero. Cohen said the money would be coming from outside Mexico and the United States. There was one other condition, Cohen added: this deal had to be done on the down-low. Huerta, as far as the outside world was concerned, would be seen as the one and only owner. With the money soon in his account, Huerta agreed.

A week later, however, Huerta began hearing around town how this gringo, Cohen, was telling everyone that he was not only an owner of the Bolero, but the only owner of the Bolero. Apparently the temptation to be the king of a club was too much to resist. Soon after, Cohen brought Brownfield down to show him his new investment. And for added cover, to make sure Kremen couldn't get his dirty hands on it, Cohen put the ownership under the name of his wife's daughter Jhuliana.

Brownfield, as usual, listened to Cohen skeptically, particularly when Cohen started bragging about how he'd already made $20,000—just in one month. "Steve," Brownfield told him, "don't talk to me about 20 grand that you made until you've got your $800,000 back. "These guys are going to screw you because they don't own that building. They're leasing the building."

But Cohen wasn't worried about them. He was too busy trying not to get screwed by Kremen: he seemed obsessed with him. Brownfield could hear him, at all hours of the day and night, calling Kremen to trash-talk him and see what he could find. "You could see in his demeanor that he was upset or he was angry," Brownfield later recalled. "He didn't want to lose that business." And Cohen told Kremen as much, when Brownfield heard him goad him on the phone about Sex.com. "You want to be in the sex business," he heard Cohen tell Kremen one day, "but you don't know anything about it."

> Dear Adult Industry, I am writing to inform you about recent changes to the Sex.Com web site. As you might know, the legal battle with Stephen Michael Cohen over the Sex.com web site ended with Cohen ordered to pay me $65 million. Additionally, the judge in the case has kept the arrest warrant open on Cohen. . . . Now that we are up and running with our new site, I wish to thank each and every member of the adult community for their support. . . . If you have any questions, give me an email at gkremen@sex.com.

It was April 2001, and as Kremen sat at his computer writing this email to the adult industry, he had reason to be grateful. After spending six long years and $2 million in legal fees chasing Cohen, he had finally won back Sex.com, the most valuable domain online. Even if he did nothing, he'd be pocketing around $5 million a year in ad sales and subscriptions, a number that, with more and more people coming online, was sure to grow. It felt like the ultimate payback, not only for refusing to give up in his pursuit of Cohen, but for having the foresight to register Sex.com years before when everyone thought he was crazy to think that property online could even exist, let alone make money. So why then wasn't he more thankful? Because until he had gotten every penny Cohen owed him, he would never rest.

His obsession with Cohen had taken over his life, especially here in the Dogpatch. While he was out chasing Cohen down south, and jetting around porn conventions, his home on 3rd Street looked more like a crack house bordello. The problem started with Crab, his local drug dealer who had essentially moved in. Ostensibly, Crab had volunteered himself to help Kremen renovate the old industrial building from a warehouse to a livable space. But a DIY carpenter high on meth isn't necessarily the right guy for the job. Walls were half demolished, paint spilled on the floor. Dust covered everything, the furniture, the steps, the bed, and Crab's burgeoning entourage of degenerates were now making Kremen's home their flophouse.

Of course, as Kremen knew, the only reason Crab had taken over was because he depended on him—for drugs. What had started with "why not?" had transformed into "where's more?" And the thing he wanted more and more of was crystal meth. In sober moments, what few there were, he couldn't believe that he had become so addicted. He felt ashamed, embarrassed, how could the son of Harriett and Norman, the Stanford MBA, the founder of online dating, become as sick as this. But then, no sooner did the thought come then it would burn away in the heat and smoke of a hit that gave him power and energy and focus, on Cohen, again.

But it wasn't just the meth that was fueling him in this next round of his fight. It was his new girlfriend: porn star Kym Wilde. A tall, tough, chestnut-haired thirty-two-year-old, Wilde had escaped an abusive childhood in New Jersey, bouncing between boarding schools and foster homes until running away to California at seventeen. While working at a hotel, a modeling agent recruited her for a shoot—which turned out to be for Larry Flynt Publishing. By twenty, she was starring in adult films.

In the decade since, she'd starred in more than two hundred adult films, many of the sadomasochistic variety (*Red Bottom Blues, Just Beat Me, Catfighting 5*, and so on). After they met through a mutual friend, Kremen'd hired her as what he called the "industrial relations specialist" for Sex.com. In essence, she was the company spokesmodel, someone who could sign autographs at conventions, pose for fans, and chat up potential advertisers.

Despite their radically different backgrounds, they liked each other from the start. Wilde was charmed by Kremen's smarts, his sweet disposition, and eccentric personality. He was the kind of guy, she thought, who needed someone to help him pick out a clean shirt and make sure his pants were zipped before a meeting. Kremen appreciated Wilde's wild streak, but also her maternal nature, how she took care of him when he needed it. They also shared a taste for drugs, but Wilde had one benefit there for him too: she was the only person who seemed capable of making sure he didn't take too much. Before long, they were dating, albeit with discretion. "I wasn't someone he was going to take home to mother," as Wilde later recalled.

Instead, she became the den mother at the Dogpatch. With Crab's help and Kremen's enthusiasm, they decided to build her an S&M dungeon in the basement where she could shoot videos and pose with Sex.com clients. And so soon enough, among the dust and debris, Crab was bringing in shackles and whips and hammering steel restraints into the wall. They installed a big wheel, where clients could be strapped, arms and legs spread, then spun in circles. They hammered a long stretching table in front, like some medieval torture device. One by one, reporters writing about Kremen's amazing case would be ushered in to the Dogpatch dungeon, given a tour by Wilde as Kremen snickered and mapped out his plan to clean up Sex.com and transform it into a mainstream site.

"It may be kind of dumb of me, but I think I can actually do it," Kremen told *Wired*. "I think I can make more money by transitioning it into a more mainstream kind of thing . . . the old site was so obnoxious people wouldn't come back," he went on. "My bet is they're going to come back here."

It wasn't bluster. He'd already severed contracts with explicit advertisers. He'd jettisoned Cohen's more garish advertisements, and replaced hard-core banners with discreet text-only links. Still, as reporters noted, a homepage pawning links to Foot Fetish and Watersports sites was hardly innocent. "It's still very definitely a porn site," as *Wired* wrote of Sex.com, "and it's probably not something that most people would call mainstream."

But then the reporters would be gone, Wilde would be sleeping, and it was just Kremen again, jacked up on speed, surfing on his computer for anything he could find on Cohen's fugitive life in Tijuana. After working with private eyes and lawyers, he'd acquired plenty of details: his address, his associates, and the rest. And that's when, one night, he got an idea. If he wasn't succeeding at smoking Cohen out of his hole, then maybe he could offer a reward for someone who could.

Kremen grabbed his keyboard. "WANTED," he typed in large bold letters, "Stephen Michael Cohen." He included Cohen's phone numbers, his email, and a description. "Stephen Michael Cohen was born on February 23, 1948 in Los Angeles, California," he wrote. "Cohen is a white male, 53 years old, with brown thinning hair, brown eyes, a gray/brown mustache, high forehead, and is approximately 5'6" to 5'8" in height. Cohen weighs approximately 220–230 pounds (perhaps more), has a florid face and nose, and is partial to wearing imprinted t-shirts." He listed Cohen's properties, the home at Rancho Santa Fe, a list of fifteen known associates. And he offered $50,000 for any information leading to Cohen's arrest.

With that, on May 30, 2001, he posted the Wanted ad on Sex .com. "I want to see to it that this doesn't happen to anyone again," Kremen told the *Register,* an internet news site. "I'll gladly pay out anyone per the legal terms and conditions, if it means putting the man who stole millions from me behind bars once and for all."

It seemed like something out a movie about the cartels. A team of bounty hunters busting down doors in Mexico. Stephen Cohen, in a panic, staring down the kidnappers who wanted to bring him back to Kremen for $50,000. But then, somehow, shots are fired, two people taking bullets, blood spilling into the battle over Sex.com.

Or, at least, that's how Cohen's attorney put it in a letter to Kremen's attorney, claiming the reward had led to a shooting in Tijuana. "What your client may have thought to be a prank has backfired," Dorband wrote. "Seven people have been arrested in Mexico attempting to abduct Mr. Cohen. Two have been shot. There is now blood on your client's hands, and I sincerely trust that you have previously advised him, based on my letter to you several weeks ago, that the reward offer should be immediately withdrawn. Your client has now exposed himself (and those acting in concert with him) to serious civil, and possibly criminal, liability. I trust you will take the appropriate action to prevent further bloodshed and even greater exposure to liability on the part of your client and his agents."

Whether it had really happened or not, no one would ultimately say. But Cohen was sure about one thing: he wasn't going to sit idle while Kremen was enlisting bounty hunters to come after him for his money, or his life. While Kremen was up in the Dogpatch working his computer to find out whatever he could on Cohen, Cohen began plotting ways to fight back. There was just one problem: he was cornered, afraid to come back into the United States since fleeing to Mexico. So whatever he was going to do had to be hatched from Tijuana. With his family out of the Rancho estate, he tried, unsuccessfully, to cancel the home insurance—again violating the order of the court against meddling with what was now

Kremen's property. He had also tried to falsify and backdate a lease to Brownfield on the guesthouse to make rendering the judgment even more tricky.

One day, Marco Moran, a young associate of Cohen's, got a call from his boss to come to his office, where he found the Toad was waiting. The twenty-seven-year-old had been working with Cohen on his various ventures, and was something of a whiz with spamming: owning blocks of network addresses, buying bandwidth, and pulling other tricks to send mass emails online in bulk.

"Gary is going to take that property," Cohen told them, referring to the house in Rancho. From what Moran could gather, as he translated their conversation, Cohen and El Sapo were hatching a scheme to make it appear as if the house was in fact owned by someone other than Cohen: a fictitious uncle of Rosey's whom Cohen named Enrique Suárez, after a real lawyer he knew in Mexico. If Cohen could prove this fake rich uncle was the real owner of Rancho, then Gary wouldn't get it because it wasn't his in the first place.

But as time passed, Cohen's plan was going nowhere—and instead he turned to more desperate measures. "At that point we already knew by him that he was losing the property," as Moran later recalled, "and he was really mad . . . he was screaming." Cohen dispatched a few other of his associates to take a trip up to Rancho. Soon after they set out for the border. They went through the grimy streets of Tijuana, across the border, up the coast, past San Diego, and into the winding roads of Rancho Santa Fe, the scent of lemons and eucalyptus in the air. They parked the car, and stepped out on the grounds—taking in the tall palms, the glistening pool, the guesthouse, the red-tiled roof. Jack Brownfield met them, acting as the contractor for what Cohen had in mind. And then they got to work.

"This is what I'm talking about right here, man," said rapper Snoop Dogg, dressed in a purple suit and black fedora, clutching his wireless microphone as two women in purple bikinis and purple boas shimmied against him. Dozens of men in pimp gear and half-dressed women crowded the stage behind him. "Keeping it real player for y'all," Snoop went on, "for the two thousand plus one at the Players Ball!"

It was June 29, 2001, at the Venetian Hotel and Casino in Las Vegas, and the biggest party in Sin City was under way. The Players Ball was a homage to the notorious party for pimps in Chicago that dated back to the early 1970s. But the players gathered this night were the rulers of a different industry: online porn. They were in town for the summer edition of Internext, the adult internet convention. It was like the quintessential inside joke. The original innovators and moguls of the digital age weren't the wannabes in Silicon Valley in the mainstream industry. The real players on the internet were the misfits and outlaws, pariahs and pornographers partying this weekend in Vegas. They had established the rules and practices that others were now racing to imitate. The party had always been theirs.

The Players Ball, started by promoter Darren "D-Money" Blatt and cohosted with his brother, the adult marketing schmoozer Kevin Blatt, had become an annual tradition. The invite-only soirees had all the trappings one would expect: blowjobs in the hallways, coke in the backrooms, porn legends Ron Jeremy and Jenna Jameson making the rounds. Rapper Ice-T and funk master George Clinton had performed at the event. The Ball culminated with the presentation of the Big Woody Award—a giant wooden penis—for biggest player of the night. This evening it was going to the featured performer, Snoop. "I'm truly honored to be holding this motherfucker right here," he said, hoisting the Woody high.

But the ultimate player of the weekend was the chubby goateed guy with dominatrix girlfriend mingling out in the crowd, Gary Kremen. And it was all because he had the biggest Woody of all, Sex.com. Everyone wanted a piece of him: the porn stars, the advertisers, the credit card processors, the Blatts. Though he'd taken Sex.com back more than six months earlier with the specific intention of keeping the site, and himself, clean, the temptations were getting the better of him.

He and Wilde had become the prom king and queen of the weekend, arm in arm, fake spanking each other for show, then slipping into the darker corners for a hit of blow.

The cocktail of sex, drugs, and money, however, couldn't keep Kremen's mind from his singular obsession: getting Cohen, and his $65 million. Yes, he was the toast of the sex industry, and already earning a half million a month. But, deep down, it meant nothing without victory over his rival. The power and paranoia sank their talons into his brain, guiding his every move against Cohen.

As the days and weeks blurred by, he worked overtime on his phone and laptop, firing off missives to Wagstaffe and digging up what he could on Cohen and his assets. Cohen's money, he found, was everywhere—a shell game with clues around the world. They subpoenaed the Bank of N. T. Butterfield of Bermuda; ATU General Trust British Virgin Islands; VP Bank Gruppe of Liechtenstein; Liechtensteinische Landesbank Aktiengesellschaft of Liechtenstein; Allgemeines Treuunternehmen of Liechtenstein; ABN Amro Bank of the Netherlands; Banco Internacional S.A. of Mexico; Banco Nacional de México S.A. of California; Banque Internationale à Luxembourg S.A. of Belgium; Rabobank International/Rabobank Utrecht of New York; and Rabo Robeco Bank S.A. of Luxembourg.

And then, finally, Wagstaffe delivered something tangible: the keys to Cohen's mansion in Rancho Santa Fe. Elated, Kremen drove down from San Francisco alone. By the time he got there, night had fallen, and with no street lamps he found himself lost in the winding roads. Tired and weary, he pulled over instead, shutting his eyes until the sun came up. When it did, it was spectacular, the golden orange hues spilling over the trees, the smell of lemons and eucalyptus in the air. Kremen had his bearings now, and continued on his way with the help of his map, until he saw it: his new home. He could hardly believe it as he parked in front of the three-car garage, and walked along the stones to the large wooden door. He could see the tennis court in the back, the kidney-shaped pool, all his.

The moment he unlocked the door and stepped onto the Mexican tile floor of the large, empty foyer, he felt his eyes search the room and his brain struggle to process the incoming data. The house had been trashed. The room was gutted: electrical sockets ripped from the walls, ceiling fans gone, wires hanging from the walls. The farther he walked, the more wreckage he registered: the door handles were ripped out, the ceiling lights gone, so too the cabinets, the shower doors, and more. In the study, the shelving looked like it had been pulled out by King Kong; in the bathrooms, the sinks were stripped down to the piping, the toilets were gone too. Then Kremen felt something underfoot—the squish of water on the carpet.

The breeze he felt wasn't coming from the air conditioner, it was coming from all the wide-open windows and doors, the screens gone, tiny black turds of four-legged visitors poppy-seeding the tile. He ran outside, to find the lemon and eucalyptus trees pulled out of the soil, the potted plants overturned, the sprinklers ripped out of the lawn, even the tennis net gone. The only furniture left outside was in broken pieces floating in the pool. As he stood there, alone in his trashed new palace, there was only one word pounding in his head: *Cohen*.

It was the first time he felt more than anger, he felt fear. Maybe it was the drugs. Maybe it was the ransacked home. But the first moment he could, Kremen went online, surfed over to Usenet, and bought himself a gun.

CHAPTER 12

COME AND GET IT

La Cruz de Loreto, a sleepy fishing village on the southwestern coast of Mexico, wasn't known far beyond the couple thousand locals. It was so remote, two hours southwest of the nearest tourist town, Puerto Vallarta, that the main place to stay in the area called itself Hotelito Desconocido, the "Unknown Hotel."

But for those in the know, the village was a hidden gem. They could spend their days on the gently lapping shore of the Peñitas de la Cruz Point beach, and take in an evening stroll through the quiet streets around the town square. Tucked between the Pacific and the Sierra Madre mountains, the area teemed with more than 150 varieties of wildlife, designated by UNESCO as an aquifer paradise. The Hotelito Desconocido drew ecotourists from around the world to stay amid the palm trees in wide, round *palafitos,* thatched cottages on stilts with their own unique names: La Pera ("the pear") or La Chalupa ("the boat").

But to Cohen, La Cruz de Loreto meant one thing and one thing only: *camarones,* shrimp. Thanks to the urging of his confidant, Jack Brownfield, he now owned 40 percent of Productora Camaronera del Mar El Ermitaño, a shrimp farm built on the estuaries here. It had been done as an appeasement to Brownfield, a bone thrown his way that meant little to Cohen but the world to his friend.

As Brownfield began work on the estuaries, he assured Cohen that his investment was going to pay back in spades. With water temperatures averaging an ideal 27 degrees Celsius, fishermen in the Cruz de Loreto Cooperativo were catching more than 10,000 pounds from the estuary every year—numbers that Brownfield expected to triple by 2003. He was getting larvae from the best sources in Mexico, he insisted, at what he said was a steal—about $5 per 1,000 larvae.

The juvenile shrimp, he went on, generally grew to about one gram in size after five weeks, safe in their predator-free nurseries, before being released into the estuary for grow-out. All in, they were sitting on a million-dollar gold mine of shrimp, easy. And for Cohen, it was all part of his plan to keep Kremen from getting the money. He was living in a big house, with a private driver, a strip club, a shrimp farm, and it was all out of Kremen's grasp. "There's nothing that Gary can do because he's never going to get the assets," Cohen told Brownfield.

But the two didn't have the luxury of talking shrimp for long. Not long after Kremen had found the house in Rancho Santa Fe trashed, the court came calling. With Cohen on the lam, the court tracked down Brownfield and asked him to send a message to his boss: if Cohen didn't return what he'd taken from the house, then the court would have no choice but to pursue his wife, Rosey, whose name was on the Rancho lease. Brownfield didn't need any more trouble with the law, and obliged. "Steve," he told Cohen, "wherever that stuff is, you've got to get it back because if you don't, they're going to arrest Rosey, you know, and that's not right because she doesn't have anything to do with this thing."

Cohen considered his options. As much as he despised Kremen, he didn't want his wife or her daughters to suffer. Even Cohen had his limits. He drew the line at his family. "Okay," he told Brownfield. "I'll give the stuff back. But they have to come and get it."

After finding the Rancho house trashed, Kremen enlisted some new help to resolve the problem: Margo Evashevski, a Stanford MBA he knew from Silicon Valley who was now working as a private eye. Bright and inquisitive with shoulder-length brown hair, Evashevski had quit the dot-com race for a higher pursuit. "What's going to make me work hard? It's not money," as she later said, "it's the truth." Evashevski had been working as a corporate investigator, gathering information for Silicon Valley companies. She enjoyed the work, solving puzzles, searching for the truth, and was known for her unrelenting work ethic.

Evashevski had met Kremen through a mutual Stanford friend, interned with him for a bit after his Match.com days, and considered him a mentor—albeit an eccentric one at that. "My peers are more conservative, but he's a risk taker," she recalled, "and he's not afraid to take a crazy idea and go ahead with it, even if everybody else is telling him that it's stupid." As friends, Kremen had a habit of calling in the middle of the night, and did so with an offer: he wanted to hire her as his private eye to dig up as many of Cohen's hidden assets as she could find.

She agreed, driving down to Rancho to meet Kremen in his new lair. What she found shocked her, even for Kremen. The bare house had turned into a fancier version of the Dogpatch, with Kym Wilde, Crab, and the others moving in alongside Kremen. Though they cleaned up the house, Kremen didn't bother to decorate, choosing instead to sleep on a mattress on the floor. Crab, the intrepid carpenter, was building a new dungeon for Kym to match the one back in the Dogpatch. Kremen let Evashevski set up her own detective's office in the house.

Kremen would be there all hours of the day and night, reveling in Evashevski's pursuit of Cohen, as she tracked down his assets, interviewed his associates, and pieced together what she could about his life. "He's a master at hiding things," she said. The way he put his things in other people's names was just the start. "He knows the exact countries to hide things in, how to move things," she went on. In all her years, she'd never encountered someone like Cohen: "He could be charming, he's very persistent when he wants something," she later recalled, but he was pathological. "He didn't seem to be able to tell the truth," as she put it. "He was so against telling the truth that he didn't even make sense. It was very difficult for me to see this skilled con man, but I knew from people that knew him back in the day that he was very good."

The more Evashevski learned of Cohen—his companies, his knack for exploiting new technologies—the more he reminded her of someone in particular. "In a lot of ways, I think he's similar to Gary," she later recalled. They both had their hands in a lot of different pots, and knew how to exploit new technologies before others. "Gary is kind of mainstream, Cohen would do everything to fuck the system," she said. "I think Cohen could have been absolutely successful if he had not always trying to skirt the law."

The law, however, was on Kremen's side when he received news that Cohen had agreed to return the property he'd stolen from the house in Rancho Santa Fe. There was just one catch: Kremen had to come get the stuff himself. Though Cohen had been ordered to turn it over, Kremen heard that Cohen had no plans to go to such efforts. So Kremen took along Evashevski and, for security, a hulking handyman named Mark to retrieve his things. They drove to the designated address, a storage shed near the Mexican border. But when they arrived, they were not alone.

Standing out front was a silver-haired Texan with narrow eyes and a Southern drawl who'd been dispatched by Cohen to hand over the goods, Jack Brownfield. Brownfield took a long look at Kremen, heavy and sweating, and thought he seemed scared. "They thought I was going to come in there with an AK-47 or something and kill everybody," he later recalled. He wasn't far off-base, regarding Kremen at least. But when Evashevski heard his name, she knew she had his number—having studied up on him in her research, where he traveled, where he shopped—and used it to her advantage. "You like to get your hair cut at Supercuts," she said, knowingly, "just north of L.A."

Brownfield eyed Evashevski. "That freaks me out a little bit," he told her. But he admired her gumption. Kremen, however, seemed a bit odd—and yet oddly familiar. His loose clothes, his sweatpants, the way he boasted about how much he was making from Sex.com. *He sounds a lot like Stephen Cohen,* thought Brownfield, who put Kremen in his place. "Look, Gary," he said, "you know, you got Sex.com, but you didn't do it, so don't be gloating to me because I know what happened." Brownfield went on, "You fell into this thing. If Steve Cohen hadn't done that and built this business, you wouldn't have got shit for that thing."

Kremen didn't care what he said, he just wanted the stuff. Brownfield opened up the storage shed, and they took a long look inside: the doorknobs, the sprinklers, the cabinets, everything Cohen had stripped was there, including the suit of armor he used to keep in the foyer. When Kremen got back to Rancho with the stuff, the first thing he did was call Cohen, whose number Evashevski had dug up, to gloat. "Hey Steve," Kremen said, in his nasally voice, "I'm sitting on the toilet in your bathroom!" Then he held up the phone so Cohen could hear him flush.

By early 2002, Cohen was growing more desperate. It was hard for him to think about anything but the one man who was angling to take his *camarones*—and everything—away: Gary Kremen. As preoccupied as Kremen had become with getting Cohen, Cohen had become consumed with keeping Kremen away. They were like two pro athletes locked in battle, Federer–Nadal, cat and mouse, catch me if you can. But just when one got a leg up on the other, the other would fight back. Cohen would be up one day, and down the next.

Despite ransacking the house in Rancho, hiding some of Sex .com riches offshore, and laundering other piles of it through the strip club and the shrimp farm, Cohen couldn't rest. His life was that of a fugitive, he couldn't even return to the country of his origin without fear of getting arrested because of the open warrant and the $65 million tab he refused to pay.

In a series of filings, he tried to get the court to drop the case. He complained of Kremen's wanted ad—that was still sending bounty hunters his way—insisted his faxes to the court were being mysteriously destroyed. As for the $65 million order, he equated it to a "death warrant," as he wrote in one declaration, saying that it violated his constitutional rights, specifically the Thirteenth Amendment—the one that abolished slavery. "Just how is the defendant expected to live?" he wrote. "How is the defendant expected to purchase the necessities of life, such as toilet paper, food, clothes and etc.? It's saying for the rest of my life that everything I own must go to Gary Kremen."

And the thing he wanted to do most was ruin every trace of anything of his before Kremen could get his paws on the goods, starting with the Rail Court property in San Ysidro, on the U.S. side of the Tijuana border, that served as his microwave internet hub. By now, Cohen was servicing thousands of customers in Tijuana, earning tens of thousands of dollars a month. As sociopathic as Cohen seemed, he really was brilliant in his own way: seeing the potential early on for dating on the net, and now as an internet provider to an underserved population. It seemed like a perfect scheme from the start. But with Kremen closing in, he could feel it all slipping away.

Dejected and prepared for the worst, Cohen called Marco Moran, his young associate, into his office one day for help. "The property in Rail Court is no longer mine," Cohen told Moran. "Gary is going to take possession of that property no matter what." If that happened, Cohen feared, Kremen would own the heart of his telecom enterprise, sending internet access to Mexico, and ruin the empire he'd so deftly engineered. He needed to buy some time, and Moran could help. First, he wanted to lease the empty lot next to the property on Rail Court. And then, Cohen said, he wanted to hire a company to put up a fence around his equipment and the Rail Court property.

Moran didn't understand what Cohen was doing, but soon went up to the property. The dusty plateau held the old shack, dotted with white microwave dishes. A metal container sat nearby with the microwave equipment inside, connected to the shack by lines of fiber. As Border Patrol cars streamed by, Moran inspected the adjacent empty field, and arranged to lease it. With the land in place, he followed Cohen's other plan, hiring a company to install a large, high fence that encircled the Rail Court property but extended far into the empty field he'd just procured. Under the sweltering sun, they sandblasted the new fence to make it look old. Then they cut the existing fence to the ground, and buried it under the dirt. As far as Moran was concerned, he was just following orders. What Cohen was up to now was anyone's guess.

Tijuana. Kremen could see the red, white, and green flag waving over rooftops just past the several lanes of cars heading toward the Border Patrol gate. But he wasn't there to wait in line to cross into Mexico. He had driven down around an hour from Rancho Santa Fe to claim his next piece of Cohen's bounty: the shack on Rail Court that Cohen was using to unlawfully beam internet access from the United States into Mexico. Kremen didn't just want the property, he wanted the business, which, according to their research, seemed to be generating tens of thousands of dollars a month.

In a parking lot near the border, he met up with Chris Jester, a mountainous, affable hacker from San Diego who'd been making his living in and around the border in the internet business. Jester had crossed paths with Cohen at one point, and helped him with the technical setup at the Rail Court property. After Margo Evashevski tracked him down, Kremen showed him the court order for the property and asked if he'd help him get inside and get it up and running again.

Jester already viewed Cohen skeptically, and was more than happy to help—considering the litany of people he claimed had been burned by Cohen already. "My understanding of everything from the horse's mouth," as Jester put it, "is that he didn't pay anyone anything. He basically created it with the plan 'oh, do this for me and I'll pay you after it makes money,' right? So he got everyone to work on that promise." Plus, the whole setup was skirting regulations, as Jester gathered, sending the microwave internet over the fence to TJ constituted "an illegal border crossing," as he put it.

Jester was savvy to the ways of the border, and had brought

along a U.S. marshal to oversee them taking the property. Supposedly, some rich family had originally owned it, planting palm trees to accentuate a house with sweeping views of Mexico. Photos existed of it with Model Ts parked outside. More recently, it had been an old used car lot at one point, Kremen learned, before Cohen had taken it over.

They turned a hard left before the Border Patrol, crossed a railroad track, and drove along a rocky dirt road alongside the tall border fence with barbed wire circling the top. As they drove slowly up the brown hill dotted with cacti, Kremen could see a run-down shack surrounded by a few dying palm trees, fiber optic cables, and, on the roof, microwave dishes pointed toward Mexico. Nearby, the word *Pacnet,* Cohen's name for his internet company, was painted on a marker. A large fence encircled the property.

Kremen eagerly bounded out of the car with his bolt cutters. With a strong squeeze he closed the jaws of the bite over the chain, feeling the satisfying snap as the links broke away before him. When he opened the shack, he and Jester were bowled over by an overwhelming smell. There was rat shit everywhere, and dead rats too, like no one had been inside for weeks, or, perhaps more likely, bothered to clean it in the heat. Covering their noses, they found a light. Racks of equipment reached to the ceiling. Jester followed a large black cable snaking through the room up to the roof, and out to a microwave dish. He and Kremen followed the direction of the dish, over the fence, over the hill, and, with binoculars, could see where it was pointed: to another dish on a black building in downtown Tijuana. They figured out that he was beaming to that location where it was distributed throughout the city—to clients including a golf course, who were paying an estimated $10,000 a month. It was the same address where Cohen had a penthouse, his mission control on the top of the black building.

Both Jester and Kremen figured that Cohen was violating all kinds of laws by smuggling internet access across the border. But seeing this in action now, the two dishes pointed to each other, the reality of what Cohen had engineered filled them with the kind of awe that two geeks like them would feel when in the presence of someone who out-geeked even them. "Motherfucker," Jester said, "this guy is pretty damn smart."

Kremen couldn't wait to shut off the service quickly enough. Back inside among the tangle of shit, dead rats, and computer equipment, Jester spotted a telephone. "I bet you if we turn off the internet somebody is gonna call," he told Kremen. Sure enough, about five minutes after they turned off the power, the phone rang. The equipment must have had some kind of alarm that notified Cohen, who was the one calling. "Hey Steve," Jester said, with a smile. Cohen assumed he was there fixing something or other. "What are you doing?" Cohen asked. "Are you working on the servers?"

"No," Jester replied. It didn't take long for Jester to start telling Cohen that he was there with Kremen when they heard the squeal of tires. They came out to see a police car and a car of Mexican men, including Marco Moran, who claimed to work for Cohen. "You vandalized the property!" one of them screamed at Kremen, pointing to the microwave dish. When the cops intervened, Kremen showed them his court order for the property. But the Mexican men did their own parceling of the area.

"This is Montano's land," one of them said, pointing to the shack, "and this is our land," he went on, pointing to the side they were on near the fence, and showing their lease. Kremen, they insisted, was trespassing here, and had no right to go inside, let alone deactivate Mr. Cohen's microwave dishes, and interrupt the internet service throughout Tijuana. "You're in a lot of trouble," one of the men told Kremen, "you just turned off Mexico."

Kremen's mind reeled. "We have a court order, this is a civil dispute," he told the cops. "This is a con man we're dealing with here." But the cops decided this was a civil matter, and left the scene. Kremen called his most recent high-powered attorney for help, Richard Idell, a San Francisco behemoth in a bow tie who'd cut his teeth representing rock promoter Bill Graham. "This is bad," Idell told him, when he heard that, assuming their map was correct, Kremen had in fact vandalized their property. "Something's wrong," Kremen said, "why would Cohen lease the property when he had all the other land?"

Kremen found out the answer after he called a master surveyor to take a look at the property. The surveyor's crew got to work, referring to maps, and using GPS gear to mark the boundaries. One of the crew worked the grounds with a metal detector, searching for any markers in the ground when his equipment picked up a signal. Brushing back the dirt, they called Kremen over and showed him what they'd found: the top of a fence, which had been cut off and covered.

Kremen realized what Cohen had done. He had extended the fence so that Kremen would be tricked into committing the crime of vandalism. Kremen was awed by Cohen's gamesmanship. He felt the worst possible feeling, the trigger that had haunted him his whole life, from his mother, from Sheldon (his childhood nemesis), and now, from Cohen. He felt outsmarted. And if Cohen outsmarted him here, he wondered, "what other things am I missing?" There was only one place he could go to find out: down the hill and across the border into Mexico.

CHAPTER 13

SEX, DRUGS, AND "CAMARONES"

"What was the look on his face?" Cohen asked. "What was his expression?"

It was shortly after Kremen's confrontation at the Rail Court property, and Cohen was in his office in Tijuana, grilling Marco Moran for details. It was the delicious payoff to his plotting, after all, how he had cut down the old fence, leased the other land, put up a new fence. He wanted Moran to tell him just how angry and frustrated Kremen had become. As Moran later recalled, "Steve obviously was very happy that he was accomplishing what he planned."

The cat and mouse game with "crazy Kremen," as Cohen had taken to calling his rival, had continued to escalate. He tried to make it a kind of entertainment, calling Kremen now and then to taunt him, or shoot the breeze. But Cohen, despite his ingenious efforts to outsmart Kremen, was getting cornered. Though he was living a comfortable life in a penthouse in Tijuana, it would have been hard to go through a day without worrying that Kremen, or a bounty hunter, might be around the corner. It wasn't just his own safety and security he had to fear, it was his wife, Rosey, and her two daughters. He couldn't even enjoy the simple pleasures of life without the possibility of it all coming to an end.

To Cohen's dismay, some of the most powerful players were aligning themselves with Kremen in his case against Network Solutions, which ruled over the domains business for years. Network Solutions was only getting bigger too, having been bought for $21 billion by another internet registration company, Verisign. The Electronic Frontier Foundation, a leading civil liberties group based in San Francisco, filed a brief on Kremen's behalf, asserting that by handing over Sex.com to Cohen, Network Solutions had recklessly abused its power. "A court has ruled that Network Solutions can screw up its monopoly on dot-com domain name management and face no consequence for its actions," EFF attorney Robin Gross said in a release. "We hope the appellate court will recognize the danger in eliminating all accountability for this key component of Internet governance."

In another blow, Cohen saw that the American Internet Registrants Association, the Washington, D.C.–based group that represented the most powerful domain owners online, had also filed a brief on Kremen's behalf. William Bode, the AIRA attorney, called the case against Kremen "fundamentally unfair" with "very substantial" stakes for anyone who wanted to establish property on the internet, whether it was a business or an individual. "There are cases like that of Kremen's in which names are lost because appropriate procedures to prevent domain name hijackings were absent," Bode went on.

On August 30, 2002, the Ninth Circuit Court of Appeals thwarted Cohen's attempt not to pay Kremen the $25 million. "In light of Cohen's status as a fugitive from justice and his egregious abuse of the litigation process, we exercise our discretion to dismiss his appeal," the three-judge panel wrote in their ruling. Fugitive from justice. Egregious abuse of litigation. The phrases sounded like what he'd been hearing his whole life. *You'll never amount to anything.* To clear his head, he liked to go to his favorite spot for lunch in Tijuana: the Costco, where he ordered himself a hot dog or two. He would sit there at his table, watching the Mexican families come and go, pushing their trolleys of oversized tuna cans and towers of paper towels. He felt at peace here in his sweatpants and polo shirt, his cell phones clipped to his belt, cutting coupons for TGI Fridays, where he might have dinner, the hot dog satisfyingly snapping between his teeth.

Crazy Kremen could go fuck himself. He wasn't going to get a penny. And Cohen wasn't going to sit around and wait any longer for him to try. After his years in the online sex business, he'd made contacts in Monte Carlo—the jewel of the French Riviera—who wanted to get into the casino business with him. He could just picture himself spending his millions there, living in a mansion by the sea, chewing on one of his favorite cigars, far from Kremen's greedy paws. Okay, he decided, *adiós* Tijuana, *bonjour* Monaco.

After infiltrating Cohen's internet shack in San Ysidro, Kremen headed down to Mexico with his private eye, Margo Evashevski. Their mission: to find as much as they could about whatever assets Cohen was hiding down there, so that he could take them back. But they couldn't just go and get him. They had to continue to build their case—and the last thing they wanted to do was tip him off to their snooping around. So they came to TJ on the down-low. This was still the art of war he was practicing, and he was on a mission. "You want to know what your opponents are doing," as he later put it.

For added security, he brought his muscle, Mark the Handyman, and the gun he'd bought online. "I listened to enough rap to know that it's better to be judged by twelve than to be carrried by six," he quipped. They were also joined by Kremen's ever-expanding team of attorneys: Tim Dillon, a San Diego–based attorney, and Alejandro Osuna, a lawyer in Tijuana. But that didn't put Evashevski any more at ease. She had learned too much about Cohen, and suspected that some of his associates in Tijuana were connected to the cartel. It didn't help when Osuna told them he'd been getting anonymous messages telling him "watch out what you're doing, we're looking at you, bad stuff can happen to you."

As they passed the sex clubs and seedy bars along Tijuana's infamous strip, Evashevski could feel her heart beating faster. For someone who was used to the dot-com scene in Silicon Valley, she felt way out of her element. They were going to meet a possible source who worked for an internet broadband provider in the city. But the meeting felt like a visit with the mob. Parked outside a modern office building, they were told to wait in their car until the employees had left after work. *We don't know what we're doing,* Evashevski thought, *this seems way over our head.*

Around 5 p.m., they were ushered inside, where a handsome young Mexican man introduced himself with a sly grin as the head of sales and marketing for the ISP. "Hello," he said, "I'm Román Caso." The second he said his name, Evashevski gulped. Caso, she knew, was one of Cohen's main associates. She had come across his name again and again in her research, linking him to Cohen's complex web of Mexican business and intrigue. Of all Cohen's associates, she considered him "target number one," as she later put it, which sent fear flashing through her brain. "I thought for sure we were going to get shot," she later recalled.

But, in fact, he wanted something else: to cooperate. There

had been an unexpected benefit to the Wanted ad that Kremen had posted online. Caso was another one who had been betrayed by Cohen's shady dealings, and wanted to help feed them information—if in part to cover his own hide. And Caso had just the place he wanted to take them. Down a busy street teeming with gringos, they came to a spot near the raucous nightclub Señor Frog's: it was the Bolero, the strip club that Cohen co-owned, and was now, Caso presumed, ready to meet the new owner, Gary Kremen.

As they went inside, Kremen looked at the ravenous customers and dancers of questionable age, and felt his stomach churn. The whole thing made him uncomfortable, even pursuing this asset at all. And the fact that he had a 9mm Glock, which he'd never had to use before, didn't make him feel any safer since he knew he probably wasn't the only one packing. His nerves didn't rest any easier when he was confronted by the men who worked there. "It's my club," Kremen said, a bit too boldly, "forty percent!"

But when word got back to Huerta, one of the owners of the Bolero, Huerta wasn't upset—like Caso, he was ready to help. Huerta never liked Cohen, and was happy to help get him out of his hair. He told them everything he could about how Cohen was getting paid, the cash deals, the credit cards. According to Dillon, a man even suggested a way they might apprehend Cohen: by having someone slip him a roofie, put him in a car, then drive him across the border. Kremen said thanks, but no thanks.

While his team began looking into taking over the Bolero business, Kremen. Margo, and Mark the Handyman hit the road for one more destination in Mexico: La Cruz de Loreto. Back in Rancho Santa Fe, he had found some of Cohen's binders tossed aside in a pile of papers, but was mystified by what he found inside: pictures of shrimp, prices of shrimp, scientific data on shrimp larvae.

Kremen took to his computer, and began searching for more information. And that's when he found it. There among the long list of websites that Cohen owned were links to Productora Camaronera del Mar El Ermitaño. *Holy camarones,* Kremen thought, *now I own a shrimp farm too.*

When they pulled up to the small fishing village, they were overcome by its beauty. The lapping waves against the large bay, the families playing on the beach. And there was a familiar person there to greet them: a lanky Texan with a weathered face, Jack Brownfield. Since meeting Brownfield at the warehouse to get back the goods from the house in Rancho Santa Fe, he had become the latest of Cohen's allies to flip, at least somewhat, their way. Brownfield still considered Cohen an ally, but he cared even more about his passion: the *camarones*. He wanted to get them into Whole Foods, to help build the local fishing community, but it was falling into disarray, plagued by theft and corruption. "You know," Brownfield said, "Steve's my friend and all, but he's not helping me with the shrimp farm."

Kremen looked out across the bay, as Brownfield began telling him about the bounty of shrimp swimming in the waters. The sun was sparkling on gentle waves. The palms swayed in the warm breeze. To calm his nerves, he'd popped a Xanax, which bathed his brain. In a lot of ways, he still felt like the fat kid from Skokie, jumping trains, stealing *Playboys*, and going on the ham radio with his dad. But life could take you as far as you could take yourself. And his life had led him here: ruler of the online porn business, owner of a mansion in one of the most expensive suburbs in America, a Tijuana strip club, a bootleg microwave internet company, and now an organic Mexican shrimp farm.

But it doesn't matter how much you have if there's something you're missing. And Kremen couldn't rest until he had what he wanted most of all, Cohen, and he could feel himself closing in. Because Cohen, he had concluded, had one fatal flaw. The more people he rubbed the wrong way, the more people Kremen could enlist to help him in his fight. "He screws over every single person around him," as Kremen later put it. "He's incapable of having a relationship with anyone."

"In my opinion," as Brownfield put it, "Steve has a list and you're on his list and depending on how desperate he is depends on whether you was gonna get burnt. You were high up enough on his list and you don't have any problems. He is a great guy, but he can use your name, use whatever he can to get whatever it is that he wants." But if Kremen would help Brownfield with the shrimp farm, he was willing to help Kremen get Cohen in return.

By 2003, the gold rush online was facing a formidable problem: piracy. It had started four years before with the release of Napster, a music-sharing service created by Northeastern university student Shawn Fanning. Before long, more robust alternatives flooded the net—Gnutella, LimeWire, BitTorrent, and others—allowing people around the world to share not just songs, but movies, games, and software. The phenomenon was threatening all the major media businesses, from the recording to the movie industries.

It was also triggering a debate about the future of media and the freedom of being a consumer. Some believed that people should be able to share what they wanted online, regardless of copyright. Others insisted that the file-sharing services themselves should not be punished for the actions of the pirates who used them. After all, people could be sharing their own music or films or games on these services as well. Why should the people suffer?

Few people, however, suspected that the latest hub for online pirates was coming out of the Middle East. Jenin, a Palestinian city in the northwestern corner of the West Bank, was mainly known for its violent clashes. The Jenin refugee camp had come under fire in April 2002 by Israel Defense Forces after it had spawned a rash of suicide bombers. But in early 2003 the Jenin refugee camp was making waves across the internet underground as the home of a controversial new file-sharing site with the retro sci-fi name Earth Station 5. A press release sent to a variety of file-sharing forums described it as a foolproof answer to the recording and motion picture industry's attack on piracy, safely tucked away far beyond their lawyers' reach.

Unlike the other services, it offered something else unique: anonymity, so that surfers could download without worrying about retribution. "Users can now freely share their music and movies online without the threat of a lawsuit from the RIAA because our technology hides each user's identity," said president Ras Kabir in a press release. "Our motto is share . . . share . . . share to your heart's content because no one can stop you."

It wasn't just a home for freedom fighters, it was a model of community. "Our group is made up of many people, Jordanians, Palestinians, Indians, Americans, Russians and Israelis," the press release went on. "Some of us are Jewish, some Christians, some Hindus and other of us are Muslim. Believe it or not, we all love and respect each other. We all work and play together. Our families on many occasions eat at the same dinner table. We trust each other and are very close friends with each other. As a group, the most important thing in our life is our children, our families and loved ones and of course our friends."

When the RIAA and MPAA threatened to sue the service for copyright violation, Earth Station 5 declared "war" on the groups, "and to make our point very clear that their governing laws and policies have absolutely no meaning to us here in Palestine, we will continue to add even more movies for FREE." The site claimed it was being used by more than nineteen million people around the world, and, while free, planned to make money eventually by offering online gambling and ads.

In a subsequent interview with *Salon* about their defiant stance, spokesperson Steve Taylor boasted that when faced with the most powerful media conglomerates in the world "we tell them to go fuck off." He went on with a laugh, "They can try [to sue us in the United States]. What are they going to do? Why don't they sue us in China? Let's say they did sue us and did win a judgment. What are they going to do, wipe their ass with it? How do they enforce it?" Matthew Oppenheim, senior vice president of legal and business affairs for the Recording Industry Association of America, responded, "Earth Station 5 is trying to get press by throwing stones at us."

But as word traveled across the internet about this enigmatically rebellious new file-sharing service, it seemed that Steve Taylor might in fact be an alias for someone else after all: Stephen M. Cohen. Though Cohen claimed to be living in Monte Carlo helping to run a casino, others suspected he had his hands in these pockets too. It would make sense. To Cohen, file-sharing was just the latest twist in the ever changing path of the internet. It was unstoppable, just as sex online had been inevitable before. Rather than fight the file-sharing services, why not try to own them and monetize them?

While investigating the site, technology reporters in Germany discovered that the notorious fugitive of Sex.com shared the same phone number and email as Taylor. When contacted, however, Taylor told *The Economist* that this was simply because they were using Cohen's Mexican internet provider for access. "Almost everything about Earth Station 5 may be fiction," the magazine concluded. "Perhaps the only certain thing is that, at least for *The Economist*, it works."

A reporter for an Israeli paper, *Yediot*, came to a similar conclusion. After contacting Taylor, the reporter was told that it was impossible to meet with Kabir, who is afraid for his safety, you probably understand, as an Israeli. Journalists are not allowed to visit ES5. Subsequently, Ras Kabir put out a press release confirming that the company had in fact hired Cohen to help their cause. "Let me make something very clear," Kabir wrote. "We offered Mr. Cohen an executive job with our company. He initially turned us down; however after several telephone calls, he finally gave in and agreed to help us in the capacity of a consultant. We now have Mr. Cohen's permission to disclose his identity."

But this did not allay the doubts. In fact, when *The Washington Post* sent a reporter to investigate, the story unraveled further. A spokesperson for the Palestinian territories phone company had no record of Earth Station 5, nor did the Palestinian Information Technology Association. The claimed location seemed to be some sort of ruse for reasons unknown. "I've never heard of the company, and I should have heard of it," said Yahya Salqan, general secretary of the association. When the reporter began asking around Jenin if anyone had heard of the company, not only did no one know them—but they found it comical. As he wrote, "Questions about its founder and president, who calls himself Ras Kabir—Arabic for 'Big Head'—drew laughter."

While Cohen was running Earth Station 5, Kremen had become the life of the party. Gone were the baggy jeans and stained sweatshirt, replaced with a black leather jacket and sunglasses. Kremen sure seemed to have reason to celebrate. He was living what many guys would imagine to be the ultimate life for someone turning forty: sex, money, and the most valuable site online. Kremen was the Sex.com king, and everyone wanted a piece of him. The pornographers wanted his business. His old Stanford buddies wanted to date the porn stars. His ever-expanding team of lawyers, private eyes, and hackers wanted money to help him get Cohen. And they all wanted the hottest ticket in the business: a trip to the house at Rancho Santa Fe, the Playboy Mansion of online porn.

When *Wired* reporter Chris O'Brien paid a few visits, he found an odd mash-up of Valley moguls and sleaze masters. As DJs spun records, Kym Wilde lolled topless in a thong in a hot tub, tweaking her nipples as Stanford geeks gawked. "My God," O'Brien heard one of Kremen's friends ask him, "where did you find her?"

Kremen demurred, as he pounded a tequila shot. "You have to have a porn star on the payroll," he said. "It's all part of the image." By way of explanation to O'Brien, Kremen likened himself to Tony Soprano. "He's like a CEO," Kremen says, "a businessman who's got business problems."

Despite the parties, Kremen was having plenty of problems of his own. The sleaze and corruption of the porn business was increasingly hard for him to stomach. The violence of some content, the misogyny, the depravity. He was never a prude, but he was earnest, in ways that people didn't understand, about trying to do what was right. Fed up with the industry's inaction, he founded a group, Adult Sites Against Child Pornography, to create more means of keeping child porn off the web.

As for his own site, he doubled down on his plan to make Sex .com a health and education destination. But transforming it into what he called the "Google of porn" wasn't making anyone rich. Cohen had been earning $500,000 a month, more than double what Kremen was making now. It was like he was in denial that he was in the porn business, and, worse, failing to deliver the gobs of money that Cohen had funneled to the adult webmasters in the past. "His mission was to do something clean and not dirty," as Jonathan Silverstein, Warshavsky's former marketing director, later put it. "People thought Gary was crazy with the direction he went with the domain."

All the while, Kremen railed against Cohen. At one party, Silverstein was beckoned into Kremen's bare bedroom and handed a phone. "Jonathan!" Kremen said in his nasally yelp, "I want you to speak with someone!" He handed him the phone, and there, on the other end, was a familiar voice: Stephen Cohen. Silverstein's head spun. Kremen had a bounty on this guy, why would they be talking? "Holy fuck," Silverstein shouted, "what's up, Steve?"

Later he asked Kremen why he'd bother with this, and Kremen insisted it was all part of his plan—the more they spoke, he hoped, the more information he could glean to help him collect. "I'm trying to get to him!" But Silverstein suspected there was some other psychology at play, some kind of twisted respect between the two in this strange cat and mouse game. "No matter how much he hated Stephen," as Silverstein later recalled, "he admired Stephen."

And he couldn't quiet the chase. When he wasn't at the Rancho mansion, he was on the trail, down to Rail Court in San Ysidro, where Wilde convinced him to let her shoot a bondage video. There in the shack, she and another woman spanked each other under the microwave dishes, filming videos to pawn online. From there, it was over the border to Bolero, Kremen in his leather jacket and shades, the Clyde to Kym's Bonnie, watching her with delight as she saddled up alongside the customers to milk whatever information she could on Cohen and his whereabouts.

From there, it was a short flight to Puerto Vallarta, where they'd be met by a local security guard who'd drive them down to the shrimp farm. Kremen was as much a celebrity there now as he was among the Stanford dot-commers and online pornmasters, a gringo, it seemed, who would be good fortune to the fishermen in town. One day, the mayor held a welcoming ceremony for Kremen, gathering at the estuary among the locals. Kremen's translator filled him in on the speech. "They're welcoming you," he said, "as a new investor trying to save the town. And they're looking for you to help out the town under the scourge of drug traffickers."

The irony wasn't lost on Kremen, who was battling with addiction of his own. The drugs were fueling his frenetic behavior, shortening what was already a huge lack of patience. Kremen couldn't keep track of it anymore: the meth, the blow, the Xanax, and whatever else came his way. Those closest to him were beginning to take notice. To his friends and family, Kremen was increasingly obsessed and dark.

Even Kym Wilde, who was dabbling herself, grew concerned. He was growing increasingly paranoid, insisting she remove the batteries from her phone so they couldn't be located. She was helping him dress, eat, and he was sleeping with a baseball bat. Phil Van Munching, his old friend from Northwestern, who was now living in New York, would be talking with Kremen one minute, when he'd say something deadpan like "I got a guy to kill in Mexico," with no inflection.

"It was like he was telling me he was having a hamburger for lunch," Van Munching later recalled.

And by the speed of Kremen's voice, the way he ranted, Van Munching suspected that drugs could be at play—a concern that others who cared about Kremen began to share among themselves. "Are you taking care of yourself?" Van Munching asked.

Kremen didn't want to lie, but he didn't want to unload either. "I don't know, dude," he'd say, "this whole thing is stressing me out."

When anyone suggested he needed to go to rehab, he'd snap back. He began hiring and firing staff, lashing out unpredictably, working all hours of the night and expecting others to do the same. "I'm so efficient, you should be in rehab," he'd snap. "I can be up for five days in a row—you guys are slackers."

And no matter how much success he was having, they could see his battle with Cohen was getting the better of him. Again and again, to their shock and mystification, they'd find him, at all hours of the day and night, on the phone with Cohen, taunting, laughing, fighting, slamming down the phone, only to dial back. Kremen had been sending him checks periodically, hoping he'd cash them so that Kremen could surreptitiously find where Cohen was banking. But, as always, Cohen had his number.

One day, Kremen received a large package in the mail. When he opened it, he found something surprising inside: an inflatable female sex doll. Sticking out of the ridiculous slit between the doll's legs was a check—the same one he'd sent to Cohen to try to have him cash—and a note. "Nice try," it read.

CHAPTER 14

HAPPY BIRTHDAY

When Kremen arrived at the U.S. Ninth Circuit Court of Appeals in San Francisco on Friday, June 27, 2003, he wasn't feeling confident. He was still smarting from losing up to this point, the court's decisions echoing in his head. He and his lawyers had argued that domains are property, that domains deserve the same protection as property, and despite the fact that he'd registered the domains for free, NSI still had a contractual obligation to fulfill. Kremen's losses were sending a larger message, a missive from the brick and mortar past, that this new world online wasn't as legitimate as the one before it. Perhaps it was a bias against the domain itself, Sex .com, a denial of the power of sex online, how it made and drove the net. But no matter. This was Kremen's last shot, and the Court of Appeals was either going to finally see the light—or relegate him to darkness once and for all.

The answer finally came in a unanimous decision from the three-judge panel, as written in a colorful opinion by Judge Alex Kozinski. "'Sex on the Internet?,' they all said. '*That*'ll never make any money,'" Kozinski wrote. "But computer-geek-turned-entrepreneur Gary Kremen knew an opportunity when he saw it . . . several years before . . . hordes of eager NASDAQ day traders would turn it into the Dutch tulip craze of our times. With a quick e-mail to the domain name registrar Network Solutions, Kremen became the proud owner of Sex.com."

But as Kozinski noted, Kremen wasn't alone—and, to Kremen's delight, Kozinski didn't pull any punches in how he qualified his rival as the "con man" that he was. "Con man Stephen Cohen, meanwhile, was doing time for he, too, saw the potential of the domain name," Kozinski went on. "Kremen had gotten it first, but that was only a minor impediment for a man of Cohen's bounded integrity."

The closer Kozinski got to the conclusion, the more Kremen's heart pounded. He'd fought so hard, been through so much, only to find that Cohen would seem to always wriggle away. As a result, he wanted even more to take down NSI, to exact justice from them for giving away Sex.com to this con man in the first place. But now, at long last, the court agreed. Kozinski ruled that NSI, in fact, was responsible for damages related to Kremen's loss of Sex.com. Kremen had won.

He leapt in the air, thrusting his fists up victoriously. *That's it! It's over! I won! I was right!* He felt electric, the power of right ruling over the evil of wrong. Cohen had swindled NSI because of their own ineptitude, he thought, and now he would bear the fruit. And he knew that this was not just a win for him. It was a massive win for the future of the internet. It marked a fundamental shift for the web as a commercial platform, and afforded the kind of protection for property that had long been customary in the "real" world. "The law has always required people who are protecting property to do so," as Jim Wagstaffe, Kremen's attorney, told *Wired*, "whether you're a coat-check person or you're running a car lot, you're obligated to protect people's property. All this case does in one sense is apply those rules to electronic property."

And, as a result, the electronic world would never be the same. "This is a landmark internet decision," Wagstaffe told the BBC. "It is the first time a court has applied traditional property protections to a domain name." In the future, in other words, if anyone tried to steal or infringe on someone's property, the owners could sue to protect what was rightfully theirs.

Kremen left the courthouse on air. That was it. He had been vindicated. Cohen was a "con man," the judge said it. NSI was liable, of course they were! They had fucked up. The internet was real. Kremen was a visionary. He had seen this. He was the mad scientist in the basement, and finally given justice. It would be up to the U.S. District Court of the Northern District of California to determine how much he'd get from NSI, but given the millions Cohen had made with his property, Kremen and his legal team figured the number could be north of $10 million.

So yeah, he had some celebrating to do. And he knew just how he wanted to do it. There was a Stanford reunion coming up, and he was going to show up in style. The big winner with the porn star at his side. He called Kym Wilde. She put on a tight black latex corset, long black boots. Kremen relished showing her off to his old classmates. He went all out for his fortieth birthday that October too.

Wagstaffe would never forget bringing his wife to the nightclub in the Mission District of San Francisco for the fortieth bash. Though the battles were hardly done—Cohen still owed Kremen $65 million, and the case against NSI was unfolding—he welcomed the opportunity to celebrate with Kremen and the rest. They'd never been to a party like this. People were wearing buttons with Kremen's face and "accidental pornographer" written on top. Pornographers, Stanford grads, Dogpatch hangers-on, friends, and family crowded the floor as music pumped and drinks flowed. Topless waitresses passed Wagstaffe and his wife, revealing a four-foot-tall ice sculpture of a penis. "This is a fun party, isn't it?" Wagstaffe told his wife with a big grin.

Before long, the party was all over, and Kremen was face-planted at the Dogpatch, when his phone rang from a strange number. He picked it up only to hear a familiar voice on the other end, Cohen.

Cohen could have been in Monte Carlo. Or Tijuana. Or the West Bank. It didn't matter. Because as far as he was concerned, he could make Gary Kremen think he was anywhere. He did this by a technique called spoofing, a little hack of the telephone system which enabled him to display any location he wanted on Kremen's Caller ID. Cohen could make a call from, say, Amsterdam, and when Kremen's phone rang, Cohen could make it so that Kremen's phone showed he was getting a call from San Diego. He liked to keep Kremen guessing. He wanted Kremen to think he was in the South of France, running a casino far from his reach, whether or not he was really there at all.

The calls had become frequent of late. And they had become longer. At all hours of the day and night. Something had changed. The chase had become so all-consuming that ordinary life began to pale. They despised each other, but they needed each other. The closer Kremen got to Cohen, the more Cohen would emerge from the shadows. By now Cohen's world was growing darker. His old allies had turned on him, Marco, Jack, the rest.

Several of the places where he was hiding his Sex.com money were getting taken by Kremen: the Rancho Santa Fe mansion, the internet shack on Rail Court, the Bolero, the shrimp farm. He had also moved his money overseas, into Luxembourg and Liechtenstein. He had gotten himself wrapped up in Earth Station 5, which never really got off the ground. He had tried to luxuriate in Monte Carlo, surrounding himself in the flash of the bare breasts and casino lights. And for what? Just to end up reaching for his phone, dialing the numbers, until the familiar nasally voice of Kremen flowed from his earpiece again. Little did Cohen know, however, Kremen was recording all their calls—just in case.

With the news of NSI's loss, Cohen was ready to congratulate Kremen on his win. He had been the first to encourage Kremen to take NSI for every penny he could, and assured him that he was going to make a mint now that they were at fault. Kremen took the occasion to wax philosophically on how far they'd both come. "What are you fifty-five, fifty-three, fifty-one?" Kremen asked Cohen.

"Fifty-five," Cohen replied. "You're what, how old are you now, forty-two?

"Just forty, just had a fortieth birthday party Saturday."

"Happy birthday."

"Thank you very much."

"You know once you get past fifty, it's sort of academic."

Cohen spoke to Kremen like Kremen was a little brother, a protégé, the kid snapping at his heels. "Cohen is someone just as twisted and smart as Gary," as Wilde had told *Playboy*. "It's what Gary admires and appreciates." He had grown accustomed to these calls, they gave him a sense of power. Kremen wanted him. He was chasing him. Cohen was playing hard to get. After all these years, they'd realized they had more in common with each other than a lot of other people. Businessmen. Entrepreneurs. Hustlers. Jewish. Short attention spans. Hands in a million pots. And early to the newest technology, a fact of which Cohen never failed to remind

Kremen. "I don't know how far back you go on the internet," Cohen said, "so, when ARPANET was around."

"Steve, I was there before you were on."

"You weren't, no, you weren't," Cohen demurred. "You were in what year?"

"Um, about 1984."

"Let me tell you something, I was there in 1979 running the French Connection. Let me tell you something . . . I was the guy that helped set up the original communications that eventually became ARPANET." No matter, they respected each other's chops, the way they both understood how sex and dating—from the French Connection to Match and Sex.com—were what really drove the internet. And they bemoaned what Cohen called the "wannabes" in the sex business compared to the legends like him including *Screw* magazine publisher Al Goldstein and Reuben Sturman, the man credited with launching the modern porn business, both of whom he claimed to have met in jail. "There are so many wannabes, Stephen," Kremen agreed, "it's a fucking pain in the ass."

"Oh, it's unbelievable," Cohen agreed. "Everybody is a genius and everybody knows how to do it better."

And they both agreed that the future of business online was file-sharing. Instead of running from it in fear like the people in the music and movie industries, the better move was to monetize it and use it to drive traffic—just as Cohen was trying to do with Earth Station 5. Kremen explained how he was putting thirty-second video clips onto Kazaa, one of the file-sharing services, with a built-in Sex.com search window at the end, so viewers would click through to the site and generate advertising revenue. "People click through like mad," Kremen told Cohen. "I learned a lot from you."

On April 20, 2004, Kremen got the icing on the cake of the court's ruling against NSI: the company decided to settle with him—for a reported $15 million. A huge smile spread across Kremen's face—$15 million! Though he was never one for material things, the windfall floored him—$15 million. He felt richer than he'd ever dreamed, vindicated. He had been right all along. Sex.com was rightfully his, not Cohen's. NSI had screwed up, for reasons that would remain a mystery: perhaps just a mistake, perhaps some conspiracy, but no matter. He had won the site back, won the staggering settlement that would leave him set if not for his whole life then most of it. "I'm ecstatic that we have reached a settlement," Kremen rejoiced in a statement, "so we can put the case behind us and find peace."

In the past when he'd felt like this, he'd pack up his car and head into the desert to unwind. And that's just what he needed now, but he didn't want to go alone. So he picked up Kym Wilde. They had grown close, despite the insanity. They'd fallen for each other. He'd taken care of her, looked out for her and her family. They'd had a birthday party at Rancho for Wilde's young son, filling a room with multicolored, plastic Chuck E. Cheese balls for him and his friends all to go diving. They drove out until the city turned into sand, and the lights turned to stars. As the night fell, they did whip-its and watched the meteor shower way up high. For a minute, it was almost like normal.

But then, back in the city, it began spiraling out of control again. Kremen's paranoia was getting the better of him. He had hired a private eye to trail Wilde, convinced she was sleeping with someone he knew. He threatened to sue her for everything she had, just as he had sued everyone else who stood in his way. Finally they were in her apartment in the Mission, arguing and fighting as she was throwing back shot after shot of tequila, until she grabbed a knife and stormed inside the bathroom, locking the door behind her.

Kremen panicked, pounding on the door, telling her to come out, until he heard an awful scream and a thud, and he put all his weight against the bathroom door, falling inside, only to find Wilde there, covered in blood, her wrists slashed, the knife on the floor. And then they were in an ambulance, squealing through the streets. She was almost dead. But she wasn't. The EMT told him that she knew what she was doing. She cut her wrists the wrong way, she was making a point. And as he watched her get wheeled away, the thick white bandages around her wrists dotted with blood, he could only ask himself one thing: *How did I get myself into this mess?* And, more important, how could he get himself out before they were carting him away next?

Before long, his relationship with Wilde had run its course. And as soon as he could, he called his sister, Julie, and told her to meet him down in Rancho Santa Fe. Julie, who was now working as a photographer, had considered her brother a whiz who sometimes went too far, but she was never going to tell him what to do. Now, however, he came calling. It was time to clean up, kick the drugs once and for all, he decided, but he needed her help. Until then, he'd tried to keep his drug use quiet, not only from her but from his parents. But when she found him there in Rancho, there was no hiding it anymore. She felt a sinking feeling when she saw her brother. This was not the guy she had known. He looked strung out, unhealthy, besieged by people who only wanted to exploit him. And she was going to do whatever she could to help him get back to himself once and for all.

First, they bought one-way tickets for all the hangers-on who were causing problems—sending Crab and the rest back to San

Francisco. She tossed out all the burritos and chips, replacing them with salad and fruit and fresh water. Kremen would go on the exercise bike, sweating out the toxins, then lock himself in his guesthouse as the cravings shook his body and rattled his mind. But soon, day after day, he felt a little better, a little healthier, until one day he emerged, clear-eyed, focused. He was okay. He would be all right. He had millions of dollars, a loving family, friends. There was just one thing left to do. And then maybe finally he would have that peace after all.

CHAPTER 15

THE SAMURAI

Just before noon on June 22, 2005, a maroon Kia Optima heading north from Mexico inched behind the long line of cars at the San Ysidro station on the border. The driver and sole occupant was Stephen Cohen's twenty-one-year-old adopted daughter, Jhuliana. As the border officer was asking her the usual questions—is this your car? "yes," anything to declare? "no"—the agent's computer showed an invalid transponder on the car. Probably nothing, especially since she was in the Secure Electronic Network for Travelers Rapid Inspection lane, designated for people who had already passed background checks. Jhuliana was directed to pull over so that her car could be further inspected.

But when the secondary inspector had Jhuliana pop the trunk, the officer found ten large packages wrapped in brown packing tape inside, along with a strange soapy smell. The agents brought over a dog, Pistol, who sniffed the car and confirmed the presence of drugs inside. The packages, which had been coated in liquid detergent to cover up the smell, were filled with pot: 202 pounds in total. Jhuliana Cohen insisted she'd only been helping out a guy named Juan whom she met in a bar, and paid her $500 to smuggle the drugs across the border. But no matter, she was under arrest.

Eight days later, Kremen sued—alleging that Stephen Cohen had been using her to conceal assets. "I guess I'll end up filing a lawsuit against [Gary] because there's absolutely no merit to it," Cohen told *Adult Video News*. "I divorced my ex-wife some time back and I have very little contact with that side of the family. I understand that [Jhuliana] was arrested for drugs. But I have no knowledge of any of this kind of stuff."

But Kremen kept closing in. One day, two months later in August, Cohen was standing outside the Tijuana office building of his lawyer, Gustavo "The Toad" Cortez Carbajal, dressed in jeans and a Beverly Hills Polo Club T-shirt, two cell phones attached to his belt, when a man came jogging up to him. "Steve Cohen?" the guy said. He introduced himself as Michael Gross, a reporter for *Playboy* who was working on a story about the Sex.com case. Gross had contacted him a few days earlier, only to be told by Cohen that he was in Monte Carlo fueling up his jet. Cohen couldn't help but be surprised. "Uh," Cohen stammered, "what are you doing here?"

When Gross began questioning him about the case, Cohen regained his composure, and invited him inside the Toad's lair. As they sat inside the darkened office, "the Toad's hand gripped my shoulder, his pockmarked face inches from mine," as Gross wrote in *Playboy*. "*Mi casa es su casa*," he told him in his Mexican accent. "Please don't steal anything." Cohen deftly explained how he had come to Tijuana. "I don't live here," he said, "I live in Europe. I'm normally in Europe. Tell Kremen you saw me. No, I'd appreciate it if you didn't. I don't want my whereabouts known to him. The days between Kremen and me are totally over. Kremen spends his life on this. I don't have the time and energy."

In October, Cohen lost a permanent injunction by Kremen, allowing Kremen to now seize all Cohen's remaining assets in the United States, from his mail to his cars, his computers to his bank accounts, all eight of them. The list read like an inventory of Cohen's heart, mind, and soul, a garage of his personal obsessions: 4 Spools of Belden Antenna Wire, 1 Lynx 35" Propane Gas Grill, 2 Gabriel Electronics HE4 220-3 DNCA Antennas.

If that wasn't enough to make Cohen's head spin, Kremen even got all his domain names, more than seventy of them, banal and lascivious. He got Bajatel.com, Ezhotmail.com, nudistcamp.com, 4fuck.info, 4fuck.net. He had letsfuckandsuck.com, love2fuck .net, wantaeatpussy.com. Kremen even won the domains for his old businesses: his strip club, the Bolero—bolerotijuana.com, Boleromensclub.com—his loan company, MexicoLending.com, and every iteration of Earth Station 5—.com, .net, .org.

Not long ago, it seemed, he had everything: the French Connection, The Club, Sex.com, a wife and children, a mansion in the lemon groves. One by one they went away, and now Kremen was taking more. But he didn't have everything. Cohen could still wake up in his penthouse and head to the Costco for a hot dog or two. No one could stop him from being a free man. When it was time to renew his residency work permit, which he was required to do since he was no longer married to Rosey, he didn't think anything of taking care of it on his own—and save the $100 it would cost to hire a lawyer. And so, on October 27, 2005, Cohen headed over to the immigration office at the border to file the paperwork himself.

But when he arrived, there was someone there to greet him: U.S. marshal Don Vazquez. Cohen, after four years of living as a fugitive, was finally put under arrest. "I was at the border waiting for him," as Vazquez admitted later. "He tried to talk his way out of it. He was trying very hard to figure out a way not to be expelled from the Republic of Mexico." Vazquez said that his office had been tipped off by an anonymous source to be on the lookout for someone fitting Cohen's description. But there was little doubt in Cohen's mind about who, in fact, was the source of his downfall: Kremen.

Downward dog was the toughest. Head down, arms outstretched, butt in the air, legs back, blood rushing to his head like a dam release setting his dark hair on end. But Kremen was determined, as he did his morning yoga with his private instructor at Rancho Santa Fe. Then it was a quick swig of his green health shake, whipped up by his sister, Julie, and a round of weights with his personal trainer. The drugs were gone from his system. His mind was as clear as the sky shining down on the pool.

Showered, fresh and clean, he trimmed his goatee, slipped on his favorite jeans, his new brown shoes, a clean yellow T-shirt, and sat on the edge of the fountain by his pool, as a photographer shot him, head tilted, a soft smile. The photographer and writer were from the *San Diego Jewish Journal*, and were there to cover the town's most famous internet antihero's latest venture: offering a $25,000 reward for anyone who could set him up with a nice Jewish woman he could marry. And if one of the relationships lasted for ten dates, he'd donate $1,000 to a Jewish charity.

As he showed off the framed articles about Match.com in his wood-paneled office, he prided himself on his achievement. "I'd like to think I've made a great contribution to the planet, I've created more love than anybody else," he said, but now he just wanted some love of his own. He'd had enough of the sex, drugs, and internet to know that it couldn't substitute for the real human connection of a partner. Though he was now worth millions of dollars, and was making another half million dollars a month from Sex

.com, it didn't matter. As he told the reporter, Judd Handler, "I'm the perfect example of money not buying happiness." The story was headlined "The Loneliest Millionaire."

But, in fact, it wasn't just the lack of love that was leaving him forlorn. It was the lack of closure with Cohen. At night, he would go to a nearby Japanese restaurant that Cohen had frequented, the Samurai, and only be reminded of his epic chase. He'd spent millions on lawyers and private eyes, and for what? For this nagging feeling of being wronged and ripped off, living solemnly in a mansion like some porned-out Citizen Kane. When he celebrated his forty-first birthday on September 20, 2005, it seemed like his nearly decade long chase of Cohen had been for naught.

But then, just a few weeks later, he got a phone call from his lawyer telling him of Cohen's arrest. Kremen couldn't believe it, but it was true. Cohen had been detained at a Mexican immigration office, and, because of Judge Ware's outstanding warrant for failing to pay Kremen his initial $25 million, deported to the United States—where he was being held without bail at the Metropolitan Correctional Center in San Diego. Kremen felt so dizzy it was like a standing downward dog. But it was real. *Finally,* he thought, *I'm going to get my money.*

But appearing in federal court the day after his arrest in a rumpled white prison jumpsuit, Cohen once again cried poverty. "I don't have a lot of financial wherewithal," Cohen told the judge, who begged to differ. "I think you'll find some disagreement on that from some quarters," the judge replied.

When Tim Dillon, Kremen's attorney, spoke with Cohen shortly after his arrest, Cohen, who seemed to Dillon to be "heavily medicated and fairly relaxed," again insisted he was too poor to pay anything near the estimated $82 million, including interest, which he owed Kremen. The best he could do, he told Dillon, was settle for $100,000. Dillon knew Kremen wouldn't settle for less than a million, and told Court TV the offer was not only laughable but that, soon after, it had dropped even lower. "I understand his offer is now $75,000 and two pieces of property," Dillon said. "I think it's classic Steve. Here he is sitting in jail, and his offer is decreasing." Even Cohen's old allies couldn't lend a hand. "Anything that is being hidden is being hidden by Steve for Steve," Jack Brownfield told Court TV. "I don't know anything about pornography. You want to know something about shrimp, come talk to me."

But Cohen wouldn't have the luxury of returning home in the meantime. On November 14, Cohen was brought before Judge Ware. Kremen eagerly showed up, eyeballing Cohen as he was led inside in his orange prison garb. Kremen could see the smirk on his face as they made eye contact. As Kremen listened, Cohen played the victim again, claiming that the reason he had not shown up at previous hearings was because he had been "physically detained" in Mexico and then went to Israel for heart surgery. "I had a duty to appear," he confessed, "and I clearly did not appear."

Cohen's latest attorney, Roger Agajanian, offered to house Cohen himself, instead of keeping him behind bars. "I've known him for a long time," Agajanian said. "He has a lot of good qualities."

Ware was unmoved. As Kremen leaned forward in his seat, he listened as Ware ordered that Cohen remain behind bars until he provided a full accounting of his money so that Kremen, at last, could finally collect. It was an uncommon decision for the court to imprison a debtor, but necessary for a con man as slippery as Cohen. "Given the seriousness of the case, it's the court's intention to hold him in custody," Judge Ware said.

At long last, Kremen thought. Ten years. Countless sleepless nights. The pills, the bills, the pain, the madness. But he was the cat who finally had cornered his mouse. As Cohen was led out, he got close enough to give him one dig. "I don't think the Samurai delivers here," Kremen told him. Cohen just narrowed his eyes. "I would never give you the money," Cohen told him, "because you stole Sex.com from me, Gary."

Cohen popped the pills. He didn't know the name. It was heart medication, prescribed from a pharmacy in Mexico. He had to have the pharmacist call the jail to get the prescription sent there. He met with a doctor in prison, who told him he'd be getting ACE inhibitors to try to shrink the size of his heart. Twice a day, he'd stand in line at pill call, inching forward until the prison employee handed him the pills in a small cup, and watch him closely as he swallowed them down.

Cohen had been imprisoned enough times throughout his life to know how to be the samurai he was inside. He'd behave, make friends, play chess or cards, watch TV. There was no law library at the prison, so he'd request legal materials, which would come in by the pile for his research. One by one, he'd read the pages, brushing up on his case, preparing for his deposition with Kremen's attorneys, determined to keep their paws far from his pockets.

On the morning of December 5, 2005, he slipped on his prison orange with the words "Santa Clara Department of Corrections Main Unit" printed in black, combed what was left of his gray hair smoothly across forehead. When he was led into the room for his deposition, he sat at the side of the table, so that his hands could be handcuffed before him. He could see Kremen staring at him, hoping for a break in his armor. But he wasn't about to show any weakness. As he was being sworn in, he kept his shit-eating grin on his face and shimmied his shoulders like a little boy in detention who wasn't going to be punished. As Cohen raised his hand to begin his deposition, Kremen could see the handcuff dangling from his wrist. His thinning gray hair was swept over his forehead, and he was grinning. "Do you solemnly swear the testimony you're about to give today shall be the truth, the whole truth, and nothing but the truth?"

"Yes," he said, with a smirk, "I do."

But extracting the truth from Cohen about the location of his money felt like battling a true samurai. The more any of Kremen's attorneys pressed, the more Cohen would slip away. He called their assessment of how much he'd made at Sex.com "voodoo accounting," estimating that maybe he made $12 million at most. When they cited his advertising revenues, as much as $1.5 million a month, Cohen demurred. "You know, I bullshit," he said. "Excuse me. I BS'd a lot of people because—because the more BS you gave, the higher you were able to charge for advertising . . . which is the standard norms for most internet-related business."

And of those earnings, the money was long gone. He had blown $4 million on the house in Rancho Santa Fe, and claimed employees at his internet company, Pacnet, stole untold more. He told stories of run-ins with "goons" in Mexico who'd been sent to his office to steal his equipment, mobsters who were coming after him because they alleged he had incriminating sex tapes he was going to use for blackmail. "I had hacienda problems up the yin yang," he said, referring to his Rancho estate. As for the location of his funds, what assets were under his family members' names, again and again he said, "I don't recall."

The song and dance continued for hours, spilling over into another deposition a few weeks later. After claiming he had been spending the last year making his living by writing and selling software, he refused to indicate how much he had earned. "As you sit here today, you can't give me an estimate of how much total you believe you collected in 2005 for the sale of your software packages?" Kremen's attorney asked, dubiously.

"Without looking at the records. Unless you want me to engage in a lie."

"An estimate is a lie?"

"An estimate without a basis—without records in front of me, would be a lie. And I'm not willing to do that. I'm here to tell the truth. And I'm willing to do that, but I'm not going to play the guessing game of—just to satisfy you. You know, there are records. I will make those records available to you. But I'm not going to play the guessing game and then have you come back and say this man lied because he's—you know, $50 off from what he should be. I'm not going to do that. I'll provide you the records. I don't know what the records show."

"Do you understand the difference between an estimate and a guess, Mr. Cohen?"

"With you, no."

"Okay. Well, let me explain."

"With anybody else, yes. I mean, there's—I mean, there's so much disingenuous conduct on your part; it's just like these subpoenas that I just learned about for the first time 20 minutes ago of which we have never received copies of them. You violate the rules of court and then you play these games where you want me to—definitive answers, and I'm willing to give you the truth. I'm willing to tell you what I know, but I am not going to lie to you. And I'm not going to play the game of guessing."

"So if I understand what you said, you don't understand my version or definition of the difference between a guess and an estimate. So I'd like to give you an example and see if you can understand that. Okay?"

"Okay."

"As you sit here today, if I asked you what the length of this table is, you can assume from your experiences in the past, you can look down at the end of the table, you can look to where you are, and you can make an estimate of about how many feet long this table is; correct?"

"Possibly."

"Do you believe you can look down and make an estimate?"

"No. I don't believe—I have no concept of what the footage of this table is, and I'm not in the business of doing that. I'm not going to play a guessing game with you, sir. You know, I will tell you the straightforward truth. The answers that I know to the best of my ability. I'm not going to exaggerate. I'm not going to guess. Where I don't have records currently available to me, but they do exist, I will make them available to you. If I'm let out one week, two weeks, two years from now, I'll still make them available to you. Nothing is changed."

But as Cohen ducked the queries like a prizefighter, whenever it was intimated that he had stolen Sex.com, he punched back. "I never stole it," he said, glancing at Kremen. "He stole it. I had—let's get this real clear. I had Sex.com since 1979 on a BBS. . . . And if I had had my day in court, I could have presented enough evidence to prove this without—without a doubt. And the name Sex .com has always been mine. . . . I take great offense that you are under the belief in all good consciousness that Sex.com was ever stolen. . . . I take offense to the fact you believe I stole Sex.com. I never stole Sex.com. I am the true owner of Sex.com."

And, as it was becoming abundantly clear, no amount of hours questioning him, not even while he was chained to a table, would break him. Cohen spit back, saying they had no right to keep him confined—and if they really wanted their money, then the only solution was to let him go. "There is nothing I can do sitting in jail," he said. "I got an attorney in Mexico that keeps telling me he is coming. And yet let's be very realistic, he has not come and he probably never will. And if I were to wait for him to come, I could be 95 years old and he still wouldn't come."

Ultimately, he conceded, none of them would have been having to deal with this at all if only he and Kremen had found some way, long ago, to get along. "A lot of this could have been prevented if Mr. Kremen and I had sat down and worked out some kind of an agreement," he said, "but at that time, because of our stubbornness between him and I, there was no agreement to be worked out. And that was the problem."

Anyone would crack in jail, Kremen figured, but not Cohen. And apparently Cohen was making something of a business for himself already behind bars. Kremen got a call from a guy claiming to be Cohen's cellmate, and offering information. He told Kremen that Cohen was hiring himself out for legal services behind bars, helping inmates with the paperwork for their appeals.

With the prospects dimming of Cohen turning over the location of his money, Kremen decided that it was time for him to start expanding his business options too. And this was the time. Five years after the dot-com bubble burst, it was, as Kremen put, "frothy" again. Investors were riding a wave of bullishness since Google had gone public in the summer of 2004, and seen its stock triple. The new social network Facebook was garnering legions of members, eager to share their lives online. Established tech titans such as Amazon, Yahoo!, and eBay were also rallying back. With billions in cash and a leash on the outlandish spending before the crash, Silicon Valley, it seemed, was finally growing up. As Tom

Taulli, cofounder of Current Offerings, an independent research firm, told CNN, "from the wreckage of the bust we now have some fundamentally good companies."

There was one guy enjoying the bubble bath who wanted to help Kremen cash in with Sex.com once and for all: Mike "Zappy" Zapolin, an ebullient Deadhead and entrepreneur known as the domain-name guru. Zappy had been one of the youngest vice presidents at Bear Stearns and an infomercial magnate before seeing gold in flipping domains. He made his name purchasing Beer.com for $80,000 and Diamond.com for $300,000, selling each for more than $7 million. His company, the Internet Real Estate Group, had become the most aggressive investors in online properties—with multimillion-dollar deals for sites including Shop.com, Telephone .com, and Computer.com. And now he wanted to help Kremen flip the most valuable domain online. "Let me sell it for you," he told Kremen.

Kremen considered his options. The boom was on. Sex was still selling, but changes were afoot. The so-called tube sites—free porn video sites such as YouPorn and PornTube that allowed surfers to watch small clips for free—were cannibalizing the online adult business. He'd cleaned up, and was tired of dealing with what he called the "kooks" of the industry. Still a Stanford MBA at heart, he was feeling the entrepreneurial itch again, eager to invest in other start-ups that could do for him in 2006 what sex and dating had done a decade before, putting him on the cusp of the next wave of technology and innovation. "Fine," Kremen told Zappy, "but I need to see eight digits."

"I'll get it," Zappy replied.

Sure enough, Zappy had just the guy: another hyperkinetic domain mogul, Mike "the Man" Mann. A self-described former "juvenile delinquent" from Baltimore who ran away from home and sold Guatemalan stones to hippie shops in L.A., Mann had gotten into the domain business after selling Menus.com, which he had bought for $70 (he'd wanted to open a vegetarian restaurant) and sold in 1998 for $25,000. From there, he became one of the biggest domain speculators online with his company, BuyDomains .com—which he sold in 2005 for $80 million.

On January 20, 2006, Mann and Zappy's shell company, Escom, bought Sex.com—for $12 million in cash and $2 million in stock, the most money every paid for a domain. Escom released a statement saying they planned to transform the site into what it called, in a statement, "the market-leading adult-entertainment destination." The *Domain Name Journal*, a trade publication, heralded it as "a landmark deal. An eight-figure sale has a way of overshadowing everything in its path."

Kremen had never been one for material things, and the money wasn't changing him now. When asked how he was going to spend his windfall, he said that, in addition to funding start-ups, he was going to purchase another four thousand domains—and make a donation to IP Justice, a San Francisco–based advocacy group for intellectual property law in emerging countries. As for whether the payout was worth the epic battle against Cohen, he remained optimistic.

"You can look at a glass as half full or half empty," he told Ron Jackson, a reporter from *Domain Name Journal*, "I see it as half full. My life could have been much better without Cohen in it, or it could have been much worse. I actually learned a lot as a victim. I learned a lot about people's character and a lot about the law. My life wouldn't have been as full if I had not been a victim." Now that he had let Sex.com go, he went on, he was letting go of some of his animosity against Cohen too. "Actually, I am sorry it has led to anyone being in jail. It's very sad because it did not have to come to this," Kremen said. "I don't believe much in retribution. It doesn't get you as much as you think it does."

Despite countless hours of questioning, Cohen never buckled. As much as Kremen's lawyers poked and prodded and tried to extract where his millions were hidden, they came up empty. And by year's end, after fourteen months behind bars, there was nothing more the court could do to detain him. "Cohen has been incarcerated for more than one year, during which time Kremen has failed to locate evidence of hidden bank accounts or other assets," Ware ruled on December 4, 2006. "Under these circumstances, the only purpose of Cohen's continued incarceration would be punitive—an impermissible purpose for civil contempt sanctions."

The next day, Cohen exchanged his prison orange for his street clothes, stepped out a free man into the California sun. It was so bright and so warm, like it had risen just for him.

CHAPTER 16

ACROSS THE BORDER

The past was dirty but the future was clean. That's what Kremen told himself after he'd sold Sex.com. Soon he would part ways with the rest: selling his interest in the shrimp farm, the Bolero, the shack on Rail Court, and the mansion in Rancho Santa Fe too. He was going back to Silicon Valley with millions to bet on the next big thing.

For twenty years, he'd proven his ability to spot and invest in trends before they hit the mainstream. The domain names he once registered for free were now part of a multibillion-dollar business of internet real estate. With Match.com, he started online dating, now a $2 billion business and essential part of love lives around the world. Even the porn business on the net was still booming, despite piracy, bringing in $3 billion a year in new forms that couldn't be pirated at all, such as live web cams. And with the rise of virtual reality, fully immersive video experiences, sex was predicted to, once again, drive the adoption of new technology just as it had done for so long in the past. The original players ball was over, but new ones would always come.

Kremen hadn't always been right, he hadn't always kept his shit together, he'd pissed off his share of people, like anyone, and he'd certainly crashed—personally and professionally—along the way. But he knew the cardinal rules of being a successful serial entrepreneur. You have to keep an open mind, and open wallet, before anyone else cashed in. You have to have the passion to pursue your dream when others don't see it. You needed to obsessively fight, if and when anyone tried to take it away. And, in the end, you had to master perhaps the toughest challenge of all: knowing when to let go. Because, sometimes, the only way to start something new was to break with the old.

The opportunity he saw now was in clean energy, specifically solar. Despite the fact that people could save money with solar panels on their homes, they were put off by the high price of installation. For $50,000, the average person would rather remodel a kitchen. To Kremen, it didn't make any sense. Here was a business, solar power, expected to become a $69 billion industry within a decade, and no one had figured out how to make it truly mass-market.

If there was one thing he learned from the porn business, it was "how to get your eye"—how to grab people's attention and get them to spend money. Solar installers were failing miserably on both counts. Their presentations were unattractive, and their price points intimidating. "Installers are guys with trucks. They're guys with ladders," as Kremen told *Mother Jones*. "They're not sophisticated marketers." Buying solar panels, Kremen thought, should be as easy as buying a car: why not just create an easy way for the average person to finance going solar?

The result was Clean Power Finance, his solar financing startup. The plan was to act as, essentially, a middleman between installers and buyers. Kremen provided two innovations: sales software, which installers could use to estimate and present a customer's savings, and affordable loans for buyers once they decided to commit. He wasn't the only one who believed. In September 2011, Kremen closed $25 million in venture capital funding for the start-up, along with the additional $75 million fund from Google. Rick Needham, Google's director of green business operations, praised Kremen's "innovative and scalable model" for its potential to "lower costs and accelerate adoption of solar energy." His alma mater, Northwestern University, invited him back to teach a course on entrepreneurism at the Institute for Sustainability and Energy.

It wasn't just his business life that was blooming. It was his personal life too. The man once called the "loneliest millionaire" was starting a family. A mutual friend had introduced him to a whip-smart doctor of internal medicine who'd immigrated from Bulgaria. Kremen liked her brains, and she liked his gumption. Plus they both wanted kids and, by 2011, had two bright-eyed boys of their own.

Kremen relished his new life of domesticity, playing with the kids and strolling with them through the Stanford campus, where he had become something of a local legend. Young entrepreneurs sought him out for advice, and investments. To fuel this, he started his own investment group, the Menlo Incubator. He launched CrossCoin Ventures, to help finance start-ups in the burgeoning arena of digital cash, such as bitcoin. He was just as brash about his predictions there as he was in the early days touting online dating. "It is unlikely that the world will ever return to a 20th-century era of paper money and plastic credit cards," as he told *Fast Company* magazine. "The future is digital money on smartphones."

With his investments in new money and green energy, he immersed himself in community service to fix up the environment his kids would be inheriting. He became president of the Purissima Hills Water District in Los Altos Hills, to help bring clean drinking water to the area, and joined the California Clean Energy Jobs Act Citizens Oversight Board, helping oversee the $650 million spent on energy efficiency upgrades to schools and public buildings. His phone was ringing so often he had to create a special message to ward off anyone who might bring him down. "Hi, you've reached Gary Kremen," his chipper message went. "Please leave positive news after the beep, and negative news before the beep."

By 2014, the man who was known for battling over the dirtiest site online had become Mr. Clean. But it only took one person to rub him the wrong way before Kremen found himself slinging mud again. While in his capacity with the Purissima Hills Water District, he had asked the chair of the Santa Clara Valley Water District, Brian Schmidt, for $5,000 to fund the installation of a pipe over a local creek.

Kremen figured it would be a no-brainer, since his area of the district had put in more than a million dollars in taxes. But to Kremen's surprise, Schmidt pushed back, suggesting there were enough wealthy people in the area to finance the creek project themselves. Kremen bristled, finding Schmidt smug and out of touch. "He really angered me with the fucking arrogance," Kremen later recalled. "He wasn't doing his job." And if Schmidt wasn't going to do his job, then Kremen would run against him and do it himself.

Kremen had found his new Stephen Cohen, a rival he became obsessed with beating. It was like he needed the competition, someone he could fight against to bring out the best in himself. He declared his candidacy for the Santa Clara Valley Water District Board, forming a campaign, pouring in his own money, hiring consultants, and going door-to-door. "It was like a start-up," Kremen said, "a well-funded start-up." He spent $130,000 on polling, printed up glossy mailers, hired Bill Clinton's political consultant

Rich Robinson. "It's very unheard of for someone to spend that kind of money in a water district race," Larry Gerston, a political science professor at San Jose State, told the *Mercury News*. "I can't tell you what his reasons are."

Neither could Schmidt, who found himself not only outspent but outwitted by his unexpected rival. When word spread that Kremen was behind a mailer mocking Schmidt's plan for recycling waste water into drinking water, Schmidt had enough: firing back with the YouTube video he shot accusing Kremen of dirty tricks. He suggested that Kremen's investment and board position in WaterSmart, a company that sold software to water districts, could create problems down the road if he won. "His continued involvement with the business could create potential conflicts of interest," Schmidt said.

Kremen denied any involvement in the toilet water mailer, and stepped down from WaterSmart, offering to donate his shares if they posed a conflict. But there was one thing he had more difficulty shaking: his involvement with Sex.com. At one point, an elderly grande dame of local politics asked him to explain why he would be involved in such a venture. Kremen took a deep breath, and told her the story. "We all have shit in our past," he concluded. And she agreed, throwing her endorsement behind him.

She was far from alone. Dozens of high-profile political leaders, including the Santa Clara County sheriff Laurie Smith and a former Democratic candidate for governor, Steve Westly, rallied behind him as well. "If he wasn't qualified and was trying to buy his way in, I would be concerned," Gilroy mayor Don Gage said of Kremen. "But from what I can tell, he is sincere. He's a bright young man." On election day, the votes came in: by 937 votes, Gary Kremen won.

Schmidt, reluctantly, accepted defeat. "I'm proud of what I

did," Schmidt told the *Palo Alto Weekly*, "running it to a near draw while being outspent 22-to-1."

And as Kremen told the *Palo Alto Weekly*, he felt surprised himself. "I guess I'm an idiot to do this," Kremen said. "All I wanted was to do something about the drought and about water. I didn't think it would get so personal." The next thing he knew, he was getting sworn in as the chair of Silicon Valley's biggest water district. It was a huge responsibility, and one that, he later admitted, he hadn't fully thought through when he entered the race. But now that he was in the job, he was ready to fight.

Those who knew Kremen best had a feeling that now, with this latest rivalry put to rest, the biggest one of his life might find a way of coming back. As his lawyer Tim Dillon put it, "Steve Cohen will do something, piss somebody off, he'll make a mistake and expose himself on some assets down the road and then we'll go after him. I don't think Gary's done."

On Friday, March 8, 2013, in Sacramento, California, a telephone rang inside the Office of Administrative Hearings, the government organization that handled hearings for more than 1,500 state and local agencies. The caller identified himself as Bob Meredith, an attorney in Utah and friend of Stephen Michael Cohen.

Cohen, Meredith explained, was scheduled for a hearing in three days regarding a private investigator's license he had held since 1983 and used under the business name The Stephen M. Cohen Investigative Agency. The license was due to expire in January but the Bureau of Security and Investigative Services had found six causes for disciplinary action that would necessitate his license being suspended or revoked.

The complaint read like a greatest hits of Cohen's con jobs:

his 1972 conviction of petty theft, which landed him in county jail; the 1975 conviction of writing bad checks, which earned him another year in the slammer and five years' probation; the 1977 conviction of grand theft, false personation, and forgery; the 1978 conviction of writing more bad checks; the 1992 conviction of bankruptcy fraud, false statements, and obstruction of justice, which earned him forty-six months in prison.

And then there was his culpability in the case of *Gary Kremen v. Stephen Michael Cohen, et al.*, in which he had been found guilty of "unlawfully taking control of the domain name, Sex.com, and profiting from its use by turning it into a lucrative online porn empire." In the eight years since he had been jailed for failing to obey the court's orders, Cohen still had yet to pay a penny of the $82 million he owed Kremen. As Cohen had recently told the bureau investigator when asked about the matter, "I have no intention of satisfying the judgment and I will not pay."

As Meredith explained on the phone, however, Cohen would not be able to make the upcoming hearing on his PI license case. The night before, he said, Cohen had been rushed to a hospital in Juárez, Mexico, and was now having open heart surgery. The hearing, therefore, would have to be set for another day, assuming Cohen survived.

When Dillon heard this news from the court, however, he arched his brow. It had been Dillon and Kremen who had reported Cohen to the Bureau of Security and Investigative Services. Kremen had long been annoyed that having a PI license was giving Cohen access to databases of personal information he might use to help his cause. But after getting notified about Meredith's call, something told Dillon that Cohen was just being Cohen again.

When he did a search on attorney Robert Meredith, a name he'd recognized from the annals of Cohen's cases, he was shocked—

even by Cohen standards—to find out just how right he was. As he read on the webpage for the Starks Funeral Parlor in Salt Lake, "Bob passed away quietly on March 7, 2013"—the day before he had allegedly called to delay Cohen's hearing. This meant one of two things: he had called from the afterlife, or Cohen had called pretending to be him. The court determined the latter. Cohen, indeed, was just being Cohen again.

And he had been since walking out of the Santa Clara County Jail in California seven years before. Buoyed by his release, Cohen was confident that he had, at long last, warded off Kremen for the very last time. But he wasn't taking any chances. When he crossed back over the border to Tijuana, he knew he would likely never be coming back to live in the United States, anywhere near the reach of Kremen, again.

In the years since, Cohen got back to rebuilding his life—in his own idiosyncratic fashion. Decades before, he had been among the first to see the future of sex online. Now that he was retired from that industry, he was moving on to the next lucrative new frontier: drugs. With the skyrocketing prices of medicines in the United States, so-called e-pharmacies were estimated to reach above $100 billion in sales by 2025. Cohen got into the business with his company Medicina Mexico, and claimed to own more than 250 licensed pharmacies throughout the country. He sold hundreds of drugs for everything from allergies to gout and weight loss through his main website, meds.com.mx. The biggest sellers were erectile dysfunction drugs such as generic Viagra and Cialis. The motto on his website: "Take care of your body. It is the only place you have to live in."

And, like Kremen, he was bullish on the future of cryptocurrency—so much so that he wrote a primer on it for anyone wishing to use it to buy drugs on his site: "BITCOINS AND

DIGITAL CURRENCIES By Stephen Cohen. Filed copyright through the U.S. Library of Congress." He touted the revolutionary aspects of bitcoin: the privacy, the anonymity, the fact that it existed far beyond the reach of banks or courts or greedy tax collectors. "Your money can't be seized by Governments, banks, creditors, or on judgments," he wrote. And, just as he predicted sex and dating would drive the internet, he insisted that bitcoin was already an inevitable. "Cash will cease to exist in future," he wrote. "Digital Currency will continue to exist where cash will not."

Though his exact earnings remained a mystery, some said he was bringing in millions per month. "I never had problems with money," as Cohen put it. "Money was never an issue. Certainly isn't now, but you didn't hear that from me. I'm set for life."

When he wasn't hustling, he was settling into his senior years—"mellowing," as his old friend Jack Brownfield, who still talked with him three times a week, put it. He was back with his ex-wife, Rosey, grandparents now to Jhuliana's two children. He'd taken up making his own beef jerky. Chris Jester, who still kept in touch with Cohen, would walk into his office to see strips of meat drying on racks. Cohen still enjoyed a cheap hot dog at Costco in Tijuana, and had become a fixture at TGI Fridays, where he'd come with his coupons and eat alone at his regular booth by the window in the back.

Aging and weary from his health problems, he was in a reflective mood late one night in 2015. "I lived a very good life," Cohen said. "I had ups and downs, but I had good times, I had million-dollar homes, I had fancy cars, I had boats, I had airplanes, I've been to just about every country." And he admitted that he hadn't always achieved these things by admirable measures. "I haven't lived the perfect life," he said. "My morals haven't been exactly up to par."

At times, he felt remorseful. "Everybody has regrets," he went on, "and I certainly have regrets. If I can look back and change things, I'd do that. With the knowledge that I have today, there are many things in my life that I could have done and would be drastically different. But unfortunately by the time you really come to realize what life is all about, you're old enough that time passed you by. And that's the problem."

But no matter how much time had passed, he still hadn't been able to completely get Kremen off his mind—particularly with Kremen and Dillon still firing shots at him across the bow to collect. One day Brownfield confronted Cohen about this. "Steve," he said, "you always tell me that Gary's water under the bridge, but every time you talk to me, you're talking to me about Gary." After finding out that Kremen was now in politics, he told Brownfield he hoped that he'd be busy enough so that "now he'll just leave me alone."

It was a hot summer day in California not long ago, as Kremen stood at the gate of yesteryear: the security entrance to his old Sex .com mansion in Rancho Santa Fe.

He wore baggy jeans and a gray T-shirt from Mount Whitney, the 14,505-foot-high mountain in the Sequoia Sierra Nevada range (the highest peak he'd ascended according to his California County High Points ranking on Peakbagger.com, a site for "summit-focused hikers"). He'd driven here from Palo Alto in his Jeep on a little tour through his past. It was the first time he'd been back since he'd sold it years before, but it still seemed the same. "I put so much money into the landscaping here," he said with a sigh, as he looked across the lemon and eucalyptus trees.

Later that afternoon, he drove forty miles south to San Ysidro, where he met Chris Jester, the cowboy hacker who'd worked with Cohen and helped Kremen take over the internet hub on Rail Court. Jester had since bought the internet hub himself. "See that dish over there?" he said, pointing to a large white dish on the rocky hill, "that's serving a five-star hotel in Tijuana."

Kremen excitedly eyed the wires along the poles. "Is this the new fiber?" he asked. "Ohhh ho ho! That looks beautiful! Oh wow!"

He pulled up to the top of the hill, where the Rail Court hub baked in the sun behind a barbed wire fence marked "Danger Restricted Area." A few dying palms stood alongside. Jester motioned to the nearby hills along the border fence. "Just in the past eight months, about seven people were shot right here," he said with a laugh. "They had a big gang fight where cartel guys came through a hole over there and they chased them up here."

But Kremen couldn't be more excited to be back. He looked for the spot where Cohen had moved the fence, eyed the old shack nearby where Cohen had set up a data center. Inside a small, one-story new building nearby, his eyes widened when he saw the tall black metal racks holding the computer servers, linked together in a spaghetti of blue, red, yellow, and orange cables. "You get the fiber all the way from L.A.?" Kremen effused.

"Yeah!" Jester replied. "There's 144 strands of fiber here!" Jester later said that he solicited help from the Border Patrol to operate here by giving them free Wi-Fi, which he also is able to monitor—for laughs. "It's the funniest thing," he said. "They stay up late at night surfing porn."

At dusk, the two men headed over the border to Tijuana to see if they might find Cohen himself. Jester said Cohen had been purposely keeping a low profile around town. "He's a hard guy to read," Jester said. "If you bump into him in the street he's so inconspicuous, he's usually wearing a pair of jeans and a baggy T-shirt." He still liked to eat his hot dogs at Costco, and clip coupons for TGI Fridays. "He uses coupons all the time so that people think that he doesn't have money," said Jester, "but I know he has money."

They drove past the strip clubs and taco stands, the zebra-donkeys pulling in red-faced tourists in cheap sombreros. As night fell, cars filled the streets, Mexican disco pumping from the night-clubs, palm trees rustling in the warm breeze, until finally they came to a small strip mall with the TGI Fridays out front. "That's his booth right there," Jester said. The booth was a few tables back by the window. They saw a Mexican family behind it, enjoying their burgers and fries.

Over dinner at a nearby steak restaurant, sharing shots of local tequila, Kremen and Jester reminisced about the old times. By the time they got out, TGI Fridays was more crowded—except for Cohen's booth, which remained conspicuously empty, as if it were reserved for the man who had made it famous. Kremen eyed it silently for a few moments, then said to no one in particular, "Maybe I should just let this go." Perhaps he had finally reached Maslow's peak on the Hierarchy of Needs, and was Self-Actualized enough to move on.

Kremen wouldn't say. But the past is easier to let go when you have someone special waiting for you now. Though his marriage to the Bulgarian doctor had proved short lived, he came out of it with two boys of his own he couldn't wait to get back to playing with—taking them on hikes, and showing them his old ham radio. And the man who still claimed to have brought more love to the world than anyone else had finally found love again for himself. She was a bright-eyed, ebullient Appalachian poet getting her master's in creative writing on a fellowship at Stanford when they met on the very place he had started long ago: Match.com, where he was still listed as the founder.

The irony of meeting her online wasn't lost on Kremen. He had started the site decades ago to find true love, after all, and twenty years was better late than never. She'd read him poetry, he'd take her camping. What the future held, he didn't know. The present was enough that, at least for now, he could leave the past behind. So he climbed back into his Jeep, eased out past TGI Fridays, and headed north across the border for the next frontier, wherever it might lead.

AUTHOR'S NOTE

In 1994, when Kremen was creating Match.com, I was working at one of New York City's first internet start-ups. I had to show up to the office building with a chunk of cement in my hand. The company, SonicNet, which hoped to become the *Rolling Stone* of the nascent online world, operated out of the owner's loft in Tribeca just off the Hudson. The desolate, cobblestone streets teemed with rats—big, vicious punk rock rodents who traveled in gangs.

To make matters more complicated, some genius, or sadist, had positioned the building's garbage cans right next to the main door. Rats love garbage, especially the oozing, stinking pizza- and condom-crusted New York variety, which meant I was greeted by a frenzied cyclone of them snapping their jaws by the door whenever I arrived. I learned quickly that to make it inside unscathed, I had to find something huge on the street to throw at the door to scatter the rats away.

Such were the lengths we had to go to at the time to get online. These were still the Wild West days of the internet, before the release of the first web browser. SonicNet existed as a Bulletin Board

System. What few users we had dialed up from their modems to slowly—oh so slowly—access our album and concert reviews. It took about an hour and a half to download an Aerosmith song. My job was to convince rock stars to come down to the loft to "chat," as we now call it, online, only then there was no word for such a thing. I'd spend my days haplessly calling music publicists begging them to send an artist downtown to go on the internet. Invariably, this would be followed by a long silence on the other end of the phone, and then the question: "What's the internet?"

As a journalist and author, I've been answering variations of that question ever since. I've written about gamers and whiz kids and moguls and hackers and social media stars. I interviewed twenty-one-year-old Facebook founder Mark Zuckerberg in his one-bedroom walk-up in Palo Alto—sparsely furnished with just a guitar, an amp, a mattress on the floor, and a teapot—when he was still handing out business cards that read "I'm CEO . . . bitch." I fielded cryptic late-night calls from an elusive Australian who'd recently launched a whistleblowing site called WikiLeaks. When I asked Julian Assange his age, he demurred because of the people who'd inevitably be coming after him, telling me "why make it easy for the bastards?"

But after writing dozens of articles and a small shelfful of books, there was still one story that I hadn't told, or seen told by someone else, yet: the Wild West years of the internet, when rats swarmed the doorways, and just getting online at all was an adventure. But what story to tell? When writing a book of narrative nonfiction, it's not enough to just say "I want to write about [blank]." There has to be a story, characters, conflict, an arc, and all that literary stuff. Oh, and it needs to sustain itself over a few hundred pages or so.

I spent several years trying to figure out what the story—or at least my version of the story—of the Wild West online would be.

The process of discovery was in itself an adventure. I went to Vegas for Defcon, a hacker convention. I went to Dundee, Scotland, to meet the creator of the video game *Grand Theft Auto*. Eventually, I found myself back in Vegas for the Players Ball, the annual soiree for the purveyors of online sex. Yes, legendary porn star Ron Jeremy—aka "The Hedgehog"—was there, along with the predictable gaggle of starlets, rappers, rockers, and celebs). But, more importantly, there was something else here: a story. How did the internet, in all its facets today, grow out of this dirty, strange aquifer?

Seeking that answer, I soon found my way to Gary Kremen, the one person who, IMHO, personified the rise of the online underground more than anyone. And from Kremen, the path led to his archnemesis, Stephen Cohen (the conflict, the arc). This book is the result of all that—a decade of reporting, dozens of interviews, cross-country flights, and late nights.

Thanks to all those who chronicled this saga as it was unfolding. Kremen's lawyer Charles Carreon captured his inside experience in his memoir *The Sex.Com Chronicles*, and journalist Kieren McCarthy tracked the case in his 2007 book, *Sex.Com*. Special thanks to Jim Wagstaffe, for providing me with several boxes of court documents and videotapes, as well as Carreon and Tim Dillon for sharing their records.

Telling a story this broad means spending a lot of time chasing down and talking with sources—including Kremen, Cohen, and others too numerous to list here. Thanks to each of you for taking the time to talk with me. All the dialogue in this book comes from court documents, articles, books, and my own interviews (for the sake of brevity and keeping myself out of the story, when I write that Kremen or Cohen or others "later said" something, that often means they said so to me directly).

I'm grateful to my editor, Jofie Ferrari-Adler, for reading/commenting/being there, and Jonathan Karp, my longtime mentor, for encouraging me to pursue this story. Thanks to the many magazine editors who put me on the road. This book isn't possible without my literary agent, David McCormick, at McCormick Literary Agency. My team at the Gotham Group—Ellen Goldsmith-Vein, Eric Robinson, Shari Smiley, and all—you're the best. And to my friends and family, nothing but love.

ABOUT THE AUTHOR

David Kushner is an award-winning journalist and author. His books include *Masters of Doom: How Two Guys Created an Empire and Transformed Pop Culture; Jonny Magic & the Card Shark Kids: How a Gang of Geeks Beat the Odds and Stormed Las Vegas; Levittown: Two Families, One Tycoon, and the Fight for Civil Rights in America's Legendary Suburb; Jacked: The Outlaw Story of Grand Theft Auto;* and *Alligator Candy: A Memoir.* Kushner is also author of the graphic novel *Rise of the Dungeon Master: Gary Gygax and the Creation of D&D,* illustrated by Koren Shadmi; and the ebook, *The Bones of Marianna: A Reform School, a Terrible Secret, and a Hundred-Year Fight for Justice.* Two collections of his magazine stories are available as audiobooks, *The World's Most Dangerous Geek: And More True Hacking Stories* and *Prepare to Meet Thy Doom: And More True Gaming Stories.*

A contributing editor of *Rolling Stone,* Kushner has written for publications including *The New Yorker, Vanity Fair, Wired, The New York Times Magazine, New York, Esquire,* and *GQ,* and has been an essayist for National Public Radio. His work is featured in several "best of" anthologies: *The Best American Crime Reporting*, *The Columbia Journalism Review's Best Business Writing*, *The Best Music Writing*, and *The Best American Travel Writing*.

He is the winner of the New York Press Club award for Best Feature Reporting. His ebook *The Bones of Marianna* was selected by Amazon as a Best Digital Single of 2013. NPR named his memoir, *Alligator Candy*, one of the best books of 2016. He has taught as a Ferris Professor of Journalism at Princeton University and an adjunct professor of journalism at New York University.